D0223992

Dignity and daily bread

In the poor countries of the world women constitute a very small proportion of the organised workforce. They are usually the recipients of the most vulnerable jobs, without any security of contract, and hardly enjoy the privileges associated with wage employment. Since women are to a large extent the workers of the 'informal' sector, their contributions to the wealth creation of an economy remain invisible to policy makers. The trade unions, likewise, frequently exclude them.

Against the background of such vulnerable forms of employment, women workers have to devise their own strategies of empowerment. By doing so they often give a new direction to labour organisations, whether inside or outside the trade union movement. The informalisation of work, however, is no longer a feature only of poor countries. The introduction of new technology and changed management policies have meant that the West too is now experiencing a new trend in the flexibilisation of employment. This has resulted in a substantial growth of feminised and casualised employment, in part-time work, contract work, shift and home-based work.

Dignity and Daily Bread analyses how global economic change is affecting women internationally and focuses on their responses to these new developments. It gives in-depth accounts of the successes and difficulties in attempts to reorganise production and describes how efforts are being made to create alternative, more humane circumstances of work and daily life.

Sheila Rowbotham has a background in economic and labour history and is the author of several books including *Women's Resistance and Revolution* (1973), *Women's Consciousness, Man's World* (1973) and *Women in Movement: Feminism and Social Action* (Routledge 1993). **Swasti Mitter** is an economist who has written extensively on homework, women and technology. She has published many books including *Common Fate, Common Bond: Women in the Global Economy* (1986), and *Computer-aided Manufacturing and Women's Employment: The Clothing Industry in Four EC countries* (1992).

Dignity and daily bread

New forms of economic organising among poor women in the Third World and the First

Edited by Sheila Rowbotham and Swasti Mitter

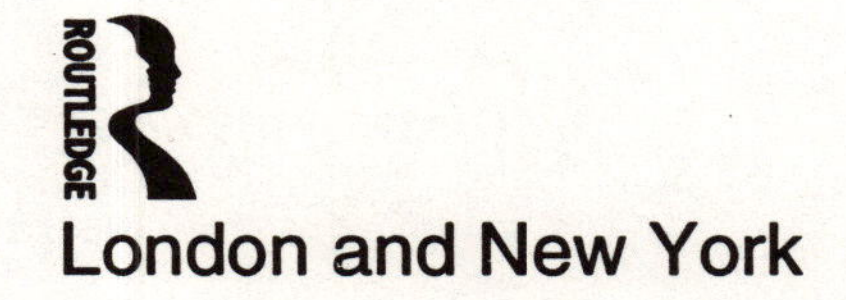

London and New York

First published 1994
by Routledge
11 New Fetter Lane, London EC4P 4EE

Simultaneously published in the USA and Canada
by Routledge
29 West 35th Street, New York, NY 10001

Typeset in Times by LaserScript, Mitcham, Surrey

Printed and bound in Great Britain by
Mackays of Chatham PLC, Chatham, Kent.

British Library Cataloguing in Publication Data
A catalogue record for this book is available from the British Library.

Library of Congress Cataloging in Publication Data
Dignity and daily bread: new forms of economic organizing among poor women in the Third World and the First/
edited by Sheila Rowbotham and Swasti Mitter.
p. cm.
Includes bibliographical references and index.
1. Women – Employment – Developing countries – Case studies. 2. Poor women – Developing countries – Case studies. 3. Women – Employment – Case studies. 4. Poor women – Case studies. I. Rowbotham, Sheila. II. Mitter, Swasti, 1939–
HD6223.D54 1994
331.4'09172'4 – dc20 93-7363
CIP

ISBN 0-415-09585-9
0-415-09586-7 (pbk)

Contents

Illustrations

FIGURES

TABLES

Contributors

Renana Jhabvala is the elected Secretary of the Self-Employed Women's Association (SEWA). She joined SEWA in 1977 after completing her MA in economics from Yale University. She has published a number of articles in journals and books on self-employed women and was awarded the Indian National Award 'Padmahri' in 1990.

Radha Kumar is a historian whose doctorate concerned women textile workers ('City Lives: Women workers in the Bombay Textile Industry, 1911–47'). She has just completed *A History of Doing*, a book on movements for women's rights and feminism in India. Currently, she is employed as Executive Director, Helsinki Citizens' Assembly, which is a coalition of civic groups in East and West Europe, the United States and Canada. In India she has been active in the women's movement (which she helped to found in the late 1970s) and the anti-communal movement, and has been involved with the Bhopal gas victims.

Swasti Mitter was educated at Calcutta University, the London School of Economics and Cambridge University. She holds a chair at the Centre for Business Research, University of Brighton, UK. She also co-ordinates the research programme on gender, technology and development at the United Nations Institute for New Technologies (UNU/INTECH), Maastricht, Holland. She has worked extensively with women workers' organisations – inside and outside trade unions – in the developed and the developing parts of the world. Her research and writings have focused on immigrant women – especially from Bangladesh – in the international division of labour. As a consultant, she has written on the areas of women and industrialisation for a wide range of international agencies, including the European Commission, UNDP, UNIFEM, ILO, UNIDO and ESCAP. Her books include *Common Fate, Common Bond* (1986) and *Computer-aided Manufacturing and Women's Employment* (1992).

Kumudhini Rosa is a sociologist. She has spent many years with women who work in the Free Trade Zones in Sri Lanka and has written extensively on this subject. As a research fellow at the Institute of Development Studies, Brighton, UK, she prepared a discussion paper about women organising in Free Trade

Zones. At present she works as a freelance consultant and researcher. She is active in women's global labour networks, working on newsletters, corresponding and speaking at meetings.

Sheila Rowbotham is a freelance writer and researcher. She is the author of many books on the history and contemporary position of women, the most recent of which are *Women in Movement* (Routledge, 1993) and *Homeworkers Worldwide* (1993). *Dignity and Daily Bread* was made possible by WIDER UNU where she worked as a consultant research adviser to the Women's Programme from 1987–91. In 1991–2 she was an associate research fellow at Birkbeck College, London University and in 1993–4 a research fellow at Manchester University. She has been active in labour organising and the women's movement for many years. Interest in economic development and global labour links grew out of her work at the Economic Policy Unit of the Greater London Council 1983–6.

Jane Tate is employed by the Yorkshire and Humberside Low Pay Unit and has been working with homeworkers in West Yorkshire for the last five years. She was active in setting up the West Yorkshire Homeworking Unit and has written reports on homework in the area as a result of the work. She has also developed extensive contacts with homeworking groups in other countries, particularly in Europe and Asia.

Silvia Tirado is an economist specialising in global economics. Active in the Mexican feminist movement since 1979, she has been involved in many workshops, meetings, conferences and radio programmes on the oppression of women. She has researched and written on the exploitative working conditions of Mexican seamstresses in the period 1975–85. Her most recent work has been on the declining economic and social hegemony of the United States. She is currently studying for her doctorate on a scholarship at the University of Munster in Germany.

Aili Mari Tripp completed her doctorate at Northwestern University in 1990. She holds a joint appointment as an assistant professor in the Department of Political Science and Women's Studies Program, University of Wisconsin-Madison. She has published numerous articles and chapters on the politics of economic liberalisation, the informal economy, responses to structural adjustment, and the new politics of participation in Africa with particular reference to Tanzania and Uganda.

Foreword

Dignity and Daily Bread is one of several studies growing from the Women's Programme of the World Institute for Development Economics Research (UNU/WIDER), a research and training centre of the United Nations University. The aim of the Women's Programme is to assert women's issues in development and also to demonstrate how gender affects economic development.

WIDER has encouraged a wide range of research exploring development issues from diverse perspectives. These have faced challenges presented by poverty, debt, environmental devastation and examined financial and economic crises along with changing forms of production and distribution. They have been concerned with the creation of policies and with the institutional machinery to ensure that these have an effect. A continuing theme has been the commitment to intervention within economies combined with a rethinking of the role of the state and the market.

Dignity and Daily Bread brings together work on the human responses to these macro-economic processes. The contributors show the importance of the general condition of the economy for women's well-being. They examine how gender relations interact with other social relationships and question an abstract separation between economy, culture and society. The specific case studies examine the impact of state policies and market forces upon poor women's lives and describe both the resistance and resourcefulness developing at the grassroots. The book chronicles a dynamism born of necessity. The editors Sheila Rowbotham and Swasti Mitter bring a combined historical and economic approach which draws out the implications of these emergent forms of organising for an extension of democratic participation in a people-centred development process. The movements described in *Dignity and Daily Bread* are crucial if economic and human resources are not to be wasted in unwieldy, inappropriate state policies which cannot be implemented or in narrowly defined market imperatives which lack any long-term social vision.

Lal Jayawardena
Director, UNU/WIDER
1993

Acknowledgements

This book developed from a seminar, 'Empowering Women in the Casualised Trades', organised by Sheila Rowbotham and Swasti Mitter in 1990 as part of the Women's Programme of the United Nations World Institute for Development Economics Research (WIDER). We are grateful to WIDER for providing the financial backing for the research and writing of the book and for the opportunity to share knowledge and ideas in the stimulating atmosphere of the Institute. The director of WIDER Lal Jayawardena's eclectic and far-ranging vision of development processes created a vital context for considering the implications of emergent forms of grassroots organisation.

In the years of preparing the seminar and papers Kumari Jayawardena, Marja-Liisa Swantz, Robin Murray, Diane Elson and Eileen Boris were important inspirational influences.

We would also like to thank the researchers at WIDER and from other academic institutes in Finland who contributed to the arguments and ideas in the book and the staff of WIDER for the administrative back-up which made our seminar possible. Thanks also to David Dougan of the Fawcett Library, Salome Mjema, assistant to Aili Mari Tripp in Tanzania, Annie Posthuma for her translation of Silvia Tirado's chapter, the typing pool and Sonia Lane who typed parts of the manuscript.

Introduction

Sheila Rowbotham and Swasti Mitter

The title of this book is drawn from a demonstration of poor women which occurred in Gujarat, India, in the summer of 1987. Women marched through the streets of Ahmedabad demanding 'Dignity and Daily Bread'. They were small vendors who wanted the right to sell their wares in the city without having to face police harassment. Mobilised by the Self-Employed Women's Association, their demonstration was not a conventional trade union protest against an employer. Nor could it be defined in terms of the community-based 'social movements' which have attracted the attention of many political theorists in the last decade. It was rather a protest of workers against deep-rooted political, social and economic forces. The women's action raised the question of the interests which determine laws and the design of cities, the allocation of space to gain a livelihood, the distribution of public resources in favour of one group or another. To women vendors of Ahmedabad, the question of access to resources appeared linked with the issue of unequal distribution of economic power; the protest, in many ways, reflected a demand for entitlement and participatory democracy on behalf of the city's poorest workers.

The vendors' strike of Ahmedabad was not an isolated event. A growing body of literature in the last decade, in the Third World as well as in the First, has brought to light a range of such mobilising experiences that contest narrowly defined concepts of class consciousness and resistance.[1]

The aim of our anthology is to consider the insights and understandings that are emerging among women workers who are unable to use traditional methods of labour organisations.

Dignity and Daily Bread documents women workers' efforts to create more democratic processes in the most varied contexts: among women in Free Trade Zones, among clothing workers in the sweatshops of Mexico City, the small entrepreneurs of Tanzania, the homeworkers of Gujarat, India, and of West Yorkshire, Britain. The accounts extend the meaning of organisation around production by revealing the interconnections of gender, race and class, and by situating work within social existence as a whole.

The main focus of the contributions is contemporary. The book nonetheless incorporates studies of earlier forms of mobilisation around women's casualised work. Both the contemporary and historical material presents perspectives which subvert the assumption that new cultural forms and their theorising originate only in the North. However, an artificial separation of 'women and development' issues in the Third World from the study of the economic and social circumstances of class, gender, race and ethnicity in the First is being increasingly challenged.[2] We have consequently included studies which span the Third World and the First.

While women's contribution is central to development, globally they rarely find employment in the regulated unionised sector, but predominate in work which evades statisticians, legislators and trade unionists. Regarded as 'atypical', the experience of millions of human beings is marginalised and their struggles to lay claim to distributive justice are frequently overlooked. Even when they are included within regulated forms of production, women's position as workers has been specifically defined by the circumstance of their gender.[3]

In *Dignity and Daily Bread* we have concentrated mainly on forms of organising and activity which have occurred outside conventional trade union structures. This is not because we dismiss the significance of more formal labour organisations. Ideally the emergent attempts would create a new synthesis with older models in which weaknesses and shortcomings could be transcended on both sides. Indeed the Indian Self-Employed Women's Association stresses that it is a union and seeks trade union support.

Dignity and Daily Bread presents only a few examples. There are innumerable other untold stories of attempts by women to lay claim to dignity and daily bread both through resistance and through networks for survival. Certainly a significant social process is occurring, but the new forms of organisation appearing among poor women producers are fragile and emergent. Their lessons for a comprehensive understanding of how to organise would necessarily be tentative and would need to draw on more extensive sources, including rural movements, which are outside the scope of this collection. *Dignity and Daily Bread* thus should be seen mainly as opening an area of enquiry for the study of new kinds of organising among poor women around production in different parts of the world.

THE PROBLEM OF DEFINITIONS

The strength of women's studies has been its habitual transgression of academic boundaries in the process of contesting the male bias embedded within conceptual frameworks. The contributors to this volume draw on economics, history, anthropology, sociology and political science. They are able to make use of a substantial body of feminist historical work on women's popular movements which, along with feminist critiques of political theories of organisation, has been contributing to a rethinking of what constitutes politics. Similarly women in development studies have made significant contributions by challenging the cultural marginalisation of women's paid and unpaid labour. While the gist of much of this work has been to dissolve the artificial separation between 'the economy' and wider social relationships, economics has remained the most impervious of disciplines to feminist critiques. For example, women's main experience of employment is persistently presented as an aberration from a male-defined norm. This has been exacerbated by a terminology which inadequately expresses women's actual experiences as workers. Because 'informal economy' or 'atypical work' suggest a deceptively distinct economic sector, separate from mainstream economic activity, we have adopted instead the term 'casualised work' or 'casualisation' as characterising much of the paid work available to women.

The phrase 'casualised work' has been used by European historians to describe the existence of seasonal or irregular employment where organisation on a continuing basis was hard to sustain. This was the general condition of men and women workers up until the mid-nineteenth century and it persisted well into the twentieth century as the situation of most women workers. In the late nineteenth century, the expression 'sweated industry' described the growth of low-paid, subcontracted, unorganised and unregulated work. It was casualised in the sense that it was barely unionised and was outside the scope of the Factory Acts.

The use of 'casualised work' as a definition is not without ambiguity or problems either. The Indian Self-Employed Women's Association prefers the phrase 'self-employed' because of the pejorative connotations of 'casual'. In western capitalism, however, 'self-employment' would not convey the predicament of many casualised women workers. Also it is possible, in the Third World as well as in the First, to be casualised by an employer either because laws are evaded or because organisation of a union is difficult or illegal. Casualisation can thus be a feature of formal employment. Indeed Radha Kumar's account of the Bombay cotton textile industry shows that this was precisely the predicament of women workers in the earlier phase of Third World industrialisation.

The modern usage of 'casualisation' expresses a pervasive shift away from regulated and unionised kinds of employment. For instance in Britain and the United States new forms of casualised work have been created by extending the contracting out of jobs, employing more homeworkers, creating franchises, expanding part-time work or passing laws which curtail state regulation of working practices.

The social implications of this process of economic transformation which accelerated during the 1980s have been much debated. Does the introduction of such flexibility mean an alternative to bureaucratic planning and the possibility of greater democratic control over work and consumption? Or is it simply a device for extracting more labour from a low-paid, vulnerable workforce? There is also considerable disagreement about even how to define these shifts in investment, production, technology, distribution and the location of industry. Is there a decisive move away from 'Fordist' mass production to a new economic order? Are we in the era of 'post-Fordism' or 'flexible specialisation', in which specialised products utilising new technology are to be made by decentralised workers? Or are we in fact seeing a more confused pattern, a new combination of employment structures in which employers favour flexibility or regulation depending on the nature of what is being produced?[4] In the midst of an unresolved debate it is clear that a break can be observed in the pattern of organisation of work in many of the richer countries and that similar trends are emerging in the newly industrialised world.

The case studies and Swasti Mitter's overview in *Dignity and Daily Bread* critically examine how economic restructuring and the emergence of a new paradigm in economic analysis affect women; for a crucial element in the identification of the obstacles which block so many women from resources and power is an assessment of the gender implications of economic policies. The specific accounts also contribute to the equally necessary task of nurturing theoretical alternatives.

WOMEN AS AGENTS OF CHANGE

Our focus is on women as social and economic protagonists, who, in their struggle for control over their working lives, carry certain potentials for differing forms of economic organisation. Lourdes Benería and Martha Roldán, in their study of homework and subcontracting in Mexico City, *The Crossroads of Class and Gender*, argued in 1987 that the disregard for agency was a weakness in the theoretical approach of much 'women in development' writing. They note a 'tendency to view women as the passive recipients of change, and as victims of forces they do not generate or control'.[5] Seeing women simply in terms of the pull of capital denies that their 'own resistance and struggles . . . while clearly subject to significant constraints due to their subordination in society, derive from a strategy of their own'.[6]

Dignity and Daily Bread seeks to extricate some of those strategies while describing the constraints within which women workers are operating. Some of these restraining factors have deep historical roots, others are defined by existing social relations and create new circumstances in which women, and indeed men, find themselves.

Correspondingly, the manner in which poor women themselves respond to the present restructuring of the economy and work provides vital clues for understanding changing social and political patterns. The accounts of women's action in this collection are specific to a point in history and geographical location, yet reveal a way of seeing that is historical and dynamic and opens up greater possibilities for human action. They provide intimations of what might be at a junction where older signposts have been razed to the ground.

Walter Benjamin's observation, that it is only as we leave something that we are able to find a name for it, is pertinent to the process of rapid social and economic transition which has dissolved the prevailing paradigms of development.[7] We are in the midst of a transformation in which the character of production, the composition of the workforce, the relation between state and economy, in fact the whole interaction between economic and social spheres, is dramatically altering. These shifts throw into question earlier assumptions about how to develop policy, plan economies and even about the process of how and by whom knowledge is constituted and defined.

FORMS OF KNOWLEDGE

The chapters in this anthology combine academic research and case studies based on the practical experience of organising. Several draw on oral testimonies as well as quantitative material. They are guided by the assumption that there is not an absolute separation between experience and the cultural processes through which human beings assimilate and reflect upon the circumstances which they encounter. Not only is theory enriched by immediate and tacit knowledge but individuals also express their ideas through their actions in the everyday; such an epistemological approach constantly seeks to link consciousness and circumstance.[8]

This orientation towards how human beings can comprehend the world has practical implications. Forms of analysis which deny agency portray society as static, while human beings are reduced to an unknowing passivity which implicitly privileges the theorist. The studies in *Dignity and Daily Bread* show in contrast how a wealth of knowing is embedded within the understandings of daily working life. Workers' tacit knowledge constitutes a potential social asset which is vital for the democratisation of power.

Human participation in economic thinking on a democratic basis provides a means of flexibly determining and fulfilling changing needs. It ensures a check on the squandering of natural and human resources, and offers a source of economic understanding which can contribute to more effective policy making. There is abundant evidence that rigid top-down planning not only fails to take human needs into account but is counter-productive.[9] The accounts in *Dignity and Daily Bread* indicate an alternative economic and social approach at a time when there is a general crisis in economic vision.

Consensual political models of development have been undermined. The efficacy of the post-war mix of state welfare and market capitalism has been challenged by advocates of the free market. Socialist state planning, likewise, has not proved to be a viable form of economic growth, or a convincing social alternative in terms of human well-being or environmental care. We are left with a vacuum in which the arbitrary criteria of the market and the interests of multinational capital predominate.

However, to say there is not a single path need not mean that no courses or patterns can be observed at all; simply that they are more complex, more open to variations, less predictable. The question of human agency is of particular significance in a period when teleological theories of historical and progressive evolution have been confuted both analytically and by actual events. Bereft of any universalising design, the social and economic protests of poor women still confront the denial of their needs, while the power and resources concentrated in the public sphere continue to be crucial to their daily lives.[10]

Lourdes Arizpe, in her foreword to the collected essays entitled *Women and Social Change in Latin America*, observes of the heterogeneous grassroots struggles for democracy and a better distribution of society's wealth, which are becoming evident,

> the fact that they occur at this moment in history – towards the end of the twentieth century – can hardly be considered a coincidence. It is highly probably that they pave the way to a different future, one whose outline we cannot yet discern.[11]

THE CONTRIBUTIONS

In the first chapter of *Dignity and Daily Bread*, Swasti Mitter outlines the changes in the organisation of production and distribution which have been transforming the lives of poor women. She shows that the impact of capital is having different effects on women in various parts of the Third World. She examines both the devastation which is part of these economic processes and the new possibilities which they present for women.

Casualised work is very much part of the economy not only in the Third World but in the First. It should not be regarded as an archaic form of production which has to be inevitably overtaken by highly capitalised units in the future. Though it has certain similarities with older forms of sweated work, the distinctive elements of the new casualisation are the use of new technology and advanced techniques of management. Swasti Mitter, in her chapter, points to the difficulty in finding a terminology which can adequately describe the emergent qualitative economic changes, and assesses the positive potential in flexibility without underplaying its negative aspects for poor women. Swasti Mitter does not endorse enthusiastically all aspects of flexibility, yet she points out that it is not inherently demeaning or unorganisable. She does not simply deplore current changes in economic structures and relations; instead she provides a broad overview which enables readers to take their bearings. The aim is to understand what is happening in order to assess the odds stacked against poor women. Her argument is that the present predicament of poor women means that their very survival depends on their capacity to create new organisational forms which enable them to earn a living and improve their lives. The experience gained by Third World poor women, in developing forms of survival and resistance, are becoming increasingly relevant to women of the First World. She demonstrates how women's networks are vital in this communication process which is overturning imperialism's legacy, the idea, in Claudette Earle's words, 'that all invention takes place in the white industrialised North'.[12] The presence of many women from the Third World within the First World economies, moreover, puts not simply gender but race and ethnicity on the agenda of women's networks. Women's mobilising experience also challenges the prevailing assumptions of who constitutes the working class and how class is to be defined.

Chapter 2 retrospectively examines women's experience in the initial phase of factory-based industrialisation in the Third World. Radha Kumar provides a comparison with modern industrialisation by showing how women were drawn into the cotton textile labour force in Bombay from the nineteenth century. By the 1920s women workers were becoming visible as active in strikes but were rarely enrolled as members of trade unions. The numbers of women in cotton textiles were to decline gradually with an overall fall in employment until World War Two. Women were not able to prevent their retrenchment and their numbers declined faster than the reduction of male workers.

The specific gender position of women affected both the manner in which they entered the cotton textile industry and the circumstances in which they were to be shed from it, often to enter an expanding casualised body of workers. Radha Kumar describes how the sexual division of labour contributed to the initial structuring of a factory-based industrial working class; industrialisation in its turn shaped and reshaped the way paid work was allocated to women.

In Bombay, casualised employment networks proliferated along with the growth of factory-based work. During the longer strikes, women reverted to these community networks for their livelihood. So there was not an absolute separation in the lives of individuals between casualised and formal systems of employment.

The Bombay cotton textile women were supported by male trade unions in disputes, but not fully integrated into the labour movement. Their specific position in the workforce and their lack of power in workers' organisations meant they could not resist retrenchment and were forced eventually into casualised forms of employment. Radha Kumar's historical account indicates how a group of women came to be deindustrialised in India.

Kumudhini Rosa's contemporary study shows that industrialisation is not a purely historical phenomenon but is still capable of evolving new forms. The creation of a new type of industrial workforce through employment in the Free Trade Zones (FTZs) is one such example. Many of the young women who entered the zones came into factory work fresh from the countryside during the 1970s and 1980s. It is consequently possible now to observe a generation of experience of the migration from one culture to another. Kumudhini Rosa describes how, initially, women workers in the FTZs saw themselves as temporary workers. Gradually, over time, their self-perception shifted, giving them a new consciousness of their permanence as workers.

There are some fascinating historical parallels. Nineteenth-century British women workers in the mills still dreamed of escape in folksongs about handsome suitors and happy weddings. In the United States, the early nineteenth-century textile mills at Lowell, Massachusetts, drew in factory workers, who came from the neighbouring farms and hoped to save in order to help to buy a farm when they married. At Lowell the employers sought to influence the attitudes of the new workforce and provided accommodation for them. This had effects similar to those that Kumudhini Rosa outlines among women in some of the FTZs. The lodging houses fostered discussion and the formation of networks which were, with the intensification of production, to contribute to a militant consciousness among what had been an educated and relatively privileged group of women workers.[13]

The links that Kumudhini Rosa describes between work and social networks are also apparent in early Japanese industry. In the late nineteenth and early twentieth centuries the Japanese Meiji textile workers included a volatile group of women and children, who were outside the formal union organisations; they too lived in allotted lodgings. Their employers sought to encourage docility, yet the women ran away, sang of their grievances and took wildcat strike action.[14]

Although much has been written about the recent establishment of FTZs, and controversy rages about whether they are beneficial to workers or not,[15] relatively little attention has been directed to the consciousness developing among women FTZ workers internationally. Indeed data upon which conclusions might be based are not yet systematically collected and researched.

Management in the zones sometimes has been prepared to pay above the current rates for women's work but has tended to be hostile to the creation of autonomous trade unions. In some countries labour rights do not apply in the FTZs. Kumudhini Rosa opens up a new area of enquiry by asking what sort of grievance has led to a range of organisational responses in Sri Lanka, Malaysia and the Philippines. These are not simply initiated by unions, but by religious and women's groups, and they span circumstances outside as well as inside the workplace.

In Chapter 4 Silvia Tirado describes how women seamstresses were employed in the small sweatshops of Mexico City, until the horrific effects of the earthquake in 1985 revealed that the factory owners were evading health and safety regulations and paying below the legal minimum to thousands of women pieceworkers. By hiring women on three-month contracts, the factory owners evaded holiday bonuses and social security legislation. Labour contracts with 'yellow unions' impeded genuine organisation and employers avoided liability by constantly dissolving and restarting companies.

The earthquake caused many slum buildings to collapse and many seamstresses were crushed to death. The garment workers and other volunteers were prevented from rescuing trapped workers by the army which cordoned off the area. Three days after the disaster women, who had formed themselves into a rescue team, broke through the cordon and managed to pull out a dozen trapped women who were still alive. On the fourth day the owners arrived with heavy equipment to retrieve machinery and raw materials but left the workers' bodies lying in the ruins.

This trauma led to the formation of the Nineteenth of September Union, an independent yet legally recognised union, amidst a glare of publicity. Silvia Tirado's account explains the significance of the union in the Mexican labour movement. She shows how the austerity measures and high prices which have led to social movements in Latin American neighbourhoods are inseparable from women's organisations as workers. They are both driven and exhausted by the combination of pressures which are being brought to bear on their lives.

This case study of a women's union, which organises around conditions of work but includes an extensive list of welfare demands, provides a useful contrast with Renana Jhabvala's case study of the Self-Employed Women's Association which follows in Chapter 5. The Self-Employed Women's Association (SEWA) is registered as a union in Gujarat, India. But it is not simply organising workers against employers like the Nineteenth of September Union. SEWA's members include small street vendors, contract workers, homeworkers and rural women workers. Many of them have no identifiable employer. Their problems include lack of access to space and cheap raw materials, lack of legislative power, lack of reasonable credit. SEWA uses co-operative and trade union strategies to bargain with municipalities and people with power over raw materials, financial institutions, property and land. Renana Jhabvala, who is the secretary of SEWA, argues that the co-operative supplements the union by providing an alternative but necessary defence for poor women workers. The co-operative also has a potential which the union alone cannot provide. It has a prefigurative element in suggesting differing economic forms in which the democratic participation of workers is a key feature.

The women workers' needs are taken as a whole and their confidence fostered as human beings in the fullest sense. SEWA arose in an area where women had been organised as textile workers and also where Gandhi had mobilised the *harijans*, those groups who had been marked by the caste system as the 'untouchables'. Gandhi's emphasis upon the immeasurable value of a sense of self-worth inspired Ela Bhatt, the founder of SEWA.

In Chapter 6, Aili Mari Tripp describes contrasting circumstances in Tanzania. Here small enterprises led by women are emerging within a state-led socialist economy confronted by runaway inflation. A wage does not provide even subsistence, so women start small businesses in order to live and to contribute to the family's income. In Tanzania, women were not traditionally breadwinners; the new phenomenon thus, while assuming a growing significance in the economy, affects gender relations even at the level of households.

Women's marginal relation to trade unions means they have not been able to defend themselves against rising unemployment through officially recognised channels. The small enterprises draw instead on personal networks among women as well as upon more formal religious and cultural associations. Aili Mari Tripp thus examines forms of productive economic activity growing within the officially recognised economy which are comparable to the social movements where personal networking is often also adapted for public purposes.

Aili Mari Tripp's study raises questions about the implications of the emergence of this small-scale 'penny capitalism' out of a rigid state socialist economy. Is it inevitable that the women must follow the pattern of sweated workers by becoming dependent upon bigger capital? Consideration of these enterprises is thus relevant not simply for Tanzanian women entrepreneurs but also for eastern Europe and for other countries which now have to think through new terms of development, amidst the problems of transition to a market economy. Can there be forms of democratic and egalitarian economic organisation of society, which avoid both the lack of flexibility of state socialism and the exploitative conditions of multinational capital's search for cheap labour?

In Chapter 7 Sheila Rowbotham provides a historical outline in which to examine this question by reassessing British attempts to organise casualised women's work in the period 1820 to 1920. She shows how assumptions about the need for state welfare and regulation of casualised work emerged out of a range of strategies which included attempts at co-operative association by both workers and philanthropists. Reliance on state intervention was part of a deeper change in attitudes towards workers, work, unions, women as mothers and reproducers and to welfare and protective legislation.

Australia and New Zealand had pioneered legislation regulating homework in the late nineteenth century partly because organised labour wanted to prevent casualised work among women and ethnic groups. In Britain, the Wages Councils were established in the early twentieth century by a combination of liberal reformers, the anti-immigration lobby, fabian socialists, feminists and progressive employers of factories like Rowntree.

The British response was part of a wider context. In Europe the effort to regulate homework produced a spate of legislation in the same period. In the United States reformers and trade unionists aimed to ban homework, a strategy which completely disregarded what women homeworkers' economic needs actually were. Eileen Boris has shown the contradictory consequences this had in the United States in several historical studies which have a contemporary significance.[16]

Working women's organisations in the 1930s New Deal combined with social reformers led by Eleanor Roosevelt, who believed firmly in protecting working-class women as mothers. This emphasis on women's biological difference was linked to state regulation which sought to improve women's conditions by welfare. This approach had some effect in the 1930s, but, in the very different economic context of the early 1980s, seemed repressive to a group of lower middle-class housewives in Vermont who wanted to earn money by knitting at home. Supported by the New Right, their grievances became a test case in the Reagan era.[17] From the 1980s, the spread of new technology was creating a new kind of homeworker; also often a relatively educated mother with children. This has led to a wider recognition that the desire for flexibility in employment was not simply imposed by backward-looking employers upon a victimised work-force but was actively sought by some workers – particularly mothers. However, the new homeworkers also face vulnerability.[18]

Jane Tate takes up some of these modern dilemmas in Chapter 8 in her case study of the West Yorkshire Homeworkers' Group. Though there are some similarities with the US, attitudes to homework in Britain have been less polarised between outright oppositionists and enthusiastic proponents. In contrast, British attempts to organise homeworkers themselves appear in the early 1970s, when community and women's groups began to try and mobilise homeworkers either into unions or as part of community politics. This was an important break with the earlier consensus of seeking state regulation. Influenced by radical ideas of democratic participation at every level of society, the groups attempted to find out what homeworkers themselves wanted.

Homeworkers' organisations first developed in Leicestershire in the 1970s, an area which in the nineteenth century had had a tradition of unionising scattered producers which persisted into modern times. West Yorkshire, which is the main focus of Jane Tate's study, had, like Ahmedabad and Bombay, an earlier history of the textile industry which included many women. Though there are several modern studies of the conditions of homeworkers, the new phase of organising is relatively untheorised. Jane Tate's examination of the organisation of homeworkers is thus breaking new ground. She shows the problems confronted at a local and national level, and also how the women organisers are operating internationally through networks and through formal channels like the International Labour Organisation (ILO). The experiences of organising are being not only accumulated locally but transmitted and compared globally as a means of putting pressure on policy makers to be responsive to homeworkers' actual needs and helping each other by publicity, exchange of information, debate about strategies. She notes that SEWA has had an important international role in Europe and Canada as well as in the Third World, where organisation of casual workers is emerging in countries like the Philippines.

These groups all stress self-organisation as a means of defining needs which then can become demands for policy both nationally and internationally. However, in order to develop this kind of grassroots participation a great deal of development work is needed. This in turn requires funding which comes from the state or, in the Third World, from aid organisations. West Yorkshire's valuable body of experience, and its recent organisational expansion to reach the various local South Asian communities, are under constant threat from a lack of resources, just as in the case of the Nineteenth of September Union. So, the long historical struggle for casualised women workers to gain access to resources has not ended, it has simply taken on new forms.

Dignity and Daily Bread describes the new contours which social action is assuming among women workers. Amidst the disintegration of older structures and paradigms, have emerged certain possibilities of political mobilisation and ideals of alternative economic organisations. Reflecting on the thought of Walter Benjamin, Hannah Arendt explains his vision of historical transformation:

> the process of decay is at the same time a process of crystallization, that in the depths of the sea, into which sinks and is dissolved what once was alive, some things 'suffer a sea change' and survive in new crystallized forms and shapes that remain immune to the elements.[19]

Synthesising old and new forms of organising will be the work of the future – that future whose outline we can as yet barely discern.

NOTES

1 See, for example, Carol Andreas, *When Women Rebel, The Rise of Popular Feminism in Peru*, Lawrence Hill & Co., Westport, Conn., 1985; Jane S. Jacquette (ed.) *The Women's Movement in Latin America, Feminism and the Transition to Democracy*, Unwin Hyman, Boston, Mass., 1989; Temma Kaplan, 'Community and Resistance in Women's Political Cultures', *Dialectical Anthropology*, 15, 1990; Jennifer Schirmer, 'Those Who Die For Life Cannot Be Called Dead', *Feminist Review*, Autumn 1989; Guida West and Rhoda Lois Blumberg (eds) *Women and Social Protest*, Oxford University Press, 1990; Swasti Mitter, *Common Fate Common Bond: Women in the Global Economy*, Pluto Press, London, 1986; Women Working Worldwide (ed.) *Common Interests: Women Organising in Global Electronics*, London, 1991; Ponna Wignaraja (ed.) *New Social Movements in the South, Empowering the People*, Zed Books, London, 1992.

2 See, Haleh Afshar, 'Women and Development: Myths and Realities; Some Introductory Notes', in Haleh Afshar (ed.) *Women, Development and Survival in the Third World*, Longman, London and New York, 1991.

3 See, for example, Pat Armstrong and Hugh Armstrong, *Theorizing Women's Work*, Garamount Press, Toronto, Canada, 1990; Diane Elson, 'Structural Adjustment: Its Effects on Women', April Brett, 'Why Gender is a Development Issue', and Peggy Antrobus, 'Women in Development', in Tina Wallace with Candida March (eds) *Changing Perceptions, Writings on Gender and Development*, Oxfam, Oxford, 1991; Sonia Kruks, Rayna Rapp and Marilyn B. Young, Introduction, *Promissory Notes, Women in the Transition to Socialism*, Monthly Review Press, New York, 1989; Jo Beall, Shiream Hassim and Alison Todas, '"A Bit On The Side", Gender Struggles in the Politics of Transformation', *Feminist Review*, 33, Autumn 1989; Gita Sen and Caren Grown, *Development Crises and Alternative Visions, Third World Women's Perspectives*, Earthscan Publications, London, 1987; see also Diane Elson (ed.) *Male Bias in the Development Process*, Manchester University Press, Manchester, 1991.

4 See Hernando de Soto, *The Other Path*, Harper & Row, New York, 1989; Alain Lipietz, *The Enchanted World*, Verso, London, 1985; Nicholas Costello, Jonathan Michie and Seamus Milne, *Beyond the Casino Economy*, Verso, London, 1989; Robin Murray, 'Benetton Britain: The New Economic Order', *Marxism Today*, November 1985; and Andrew Sayer and Richard Walker, *The New Social Economy: Reworking the Division of Labour*, Blackwell, Oxford, 1992.

5 Lourdes Benería and Martha Roldán, *The Crossroads of Class and Gender, Industrial Homework, Subcontracting and Household Dynamics in Mexico City*, University of Chicago Press, 1987, p. 8.

6 ibid.

7 See Allal Sinaceur, 'Histoire, culture et traduction', in Françoise Barret-Ducrocq, *Traduire l'Europe*, Documents Payot, Paris, 1992, p. 48.

8 See E. P. Thompson, 'The Poverty of Theory', in ed. E. P. Thompson (ed.) *The Poverty of Theory and Other Essays*, Merlin, London, 1978, p. 363.

9 See Maureen Mackintosh and Hilary Wainwright (eds) *A Taste of Power: The Politics of Local Economics*, Verso, London, 1987; Mike Cooley, *Architect or Bee? The Human Price of Technology*, Chatto & Windus, London, 1987; Swasti Mitter, *Computer-aided Manufacturing and Women's Employment: The Clothing Industry in Four EC Countries*, Springer-Verlag, London, 1992; and Michael Albert and Robin Hahnel, *Looking Forward, Participatory Economics for the Twenty-first Century*, South End Press, Boston, 1991.

10 See Jacquette, *The Women's Movement in Latin America, Feminism and the Transition to Democracy*, and Elizabeth Jelin, *Women and Social Change in Latin America*, UNRISD, Geneva, and Zed, London, 1990.

11 Lourdes Arizpe, 'Foreword: Democracy for a Small Two-Gender Planet', in Elizabeth Jelin (ed.) *Women and Social Change in Latin America*, UNRISD, Geneva, and Zed Books, London, 1990, p. xiv.
12 Claudette Earle, 'Media Concepts for Human Development in the Caribbean with Special Reference to Women', in Pat Ellis (ed.) *Women of the Caribbean*, Zed, London, 1988, p. 116.
13 See Philip S. Foner, *Women and the American Labor Movement, From the First Trade Unions to the Present*, The Free Press, Macmillan, New York, 1974, pp. 33–4; and Thomas Dublin, *Women at Work: The Transformation of Work and Community in Lowell, Massachusetts, 1826–1860*, Columbia University Press, New York, 1979.
14 See Patricia Tsurumi, 'Female Textile Workers and the Failure of Early Trade Unionism in Japan', *History Workshop, a journal of socialist and feminist historians*, 18, Autumn 1984.
15 See Diane Elson and Ruth Pearson, 'The Latest Phase of the Internationalisation of Capital and Its Implications for Women in the Third World', Institute of Development Studies, Sussex, Discussion Paper 150, 1980; and Gillian H. C. Foo and Linda Y. C. Lim, *Poverty, Ideology and Women Export Factory Workers in South-East Asia*, Macmillan Press, London, 1989.
16 See Eileen Boris, 'Regulating Industrial Homework: The Triumph of "Sacred Motherhood"', *Journal of American History*, 71(4), March 1985; and Eileen Boris and Cynthia R. Daniels (eds) *Homework: Historical and Contemporary Perspectives on Paid Labour at Home*, University of Illinois Press, Urbana and Chicago, 1989.
17 See Eileen Boris, 'Homework and Women's Rights: The Case of the Vermont Knitters, 1980–1985', *Signs: Journal of Women in Culture and Society*, 13(1), 1987.
18 See Ursula Huws, *The New Homeworkers: New Technology and the Relocation of White-Collar Work*, Low Pay Unit, London, 1984. For an outline of the discussion on flexibility in employment since the 1970s, see Ursula Huws, 'Telework: Projections', *Futures*, January/February 1991.
19 Hannah Arendt, 'Introduction, Walter Benjamin: 1892–1940', in Walter Benjamin, *Illuminations*, Fontana, London, 1973, p. 51.

Chapter 1

On organising women in casualised work

A global overview

Swasti Mitter

ABSTRACT

During the 1980s new structures in the organisation of production and new global patterns of investment appeared. These have had significant consequences upon the lives of poor women in the Third World and in the First. The employers' drive for flexibility which new technology makes feasible has intensified pressure upon the working conditions of women. Women, significant in the emerging labour forces, are vulnerable both as unskilled cheap labour and because they tend to be marginal to existing forms of trade unionism. Structural changes in patterns of work and investment makes mobilisation through existing trade union models difficult, either because workers are in small units or because multinational capital, helped by governments who need investment, is able to prevent organisation.

Swasti Mitter describes how in response to these circumstances various efforts to find appropriate forms of resisting casualisation are beginning to emerge. These include attempts to organise women at work as well as efforts to gain access to social resources through action in communities.

These tentative strategies of survival and rebellion raise vital questions for the international trade union movement and for economic policy.

THE GROWTH OF CASUALISED WORK

The characteristics of casualised work

In the developed as well as in the developing world, women find it difficult to get core jobs in the mainstream economy that are adequately protected by a country's employment and labour legislation. The organised or the official sector – often described as the formal sector – is dominated by large-scale companies where it is easy to monitor the implementation of labour laws. The consequent obligation of the employers to pay women specific benefits, such as maternity benefits or child care in some countries, pushes up the cost of hiring women workers. Women's responsibilities at home, towards children and older relatives, increase their absenteeism at the place of work. It is not difficult to understand, therefore, why employers prefer to hire men in permanent and protected jobs. In contrast, women when hired in the mainstream economy are recruited mainly on flexible and terminable contracts.

The exclusion of women from the mainstream economy is particularly visible in the poorer parts of the world. In these countries poor women survive by setting up on their own as minuscule entrepreneurs. Alternatively, they find jobs in the unorganised sector, where employers are either exempt from the relevant legislation or cannot be monitored for their non-compliance with it. In the unorganised sector, it is easy to hire and fire workers, and here, significantly, women are often recruited in preference to men.

The unorganised sector, by definition, does not lend itself to statistical quantification. Women's productive activities thus remain unaccounted for in the standard measurement procedures of the Gross National Products.[1] As a result, a vast majority of women, in spite of their active participation in the sphere of paid employment, are seen as economically unproductive. In the policy documents of national and international planners, women at best get mentioned as the workers of the informal sector, a sector that is perceived to exist only as an appendage to the mainstream formal sector.

Despite the frequent use of the term, the 'informal sector' is not easy to define. Academics and policy makers have used the concept differently and in different contexts. In Italy, for example, the term is used interchangeably with *lavoro nero*, the black economy.[2] To the Greens in the UK, it signifies those activities needed for human well-being and for which no financial transactions are involved.[3]

The above-mentioned definitions or descriptions of the informal sector are not without problems. Equating the informal sector with the black economy, as in Italy, implies the illegal or semi-criminal nature of this specific sector. The definition of the Green Party, on the other hand, evokes an alternative craft-based form of production geared essentially to non-market consumption; the Green definition also idealises certain aspects of poor women's experiences of unpaid labour. In development literature, however, the definition most commonly used is that of the International Labour Organisation (ILO), which typifies the informal sector as consisting of minuscule units and labour-intensive methods of production. This definition or description has the optimistic certainties of the early 1970s, which saw in the labour-intensive forms of production of the informal sector a solution to the staggering unemployment problems of the poorer parts of the world.[4]

To those who were less enthusiastic about the sector even in the 1970s, the informal economy was generally seen as a residual sector which was, in the course of time, to be overtaken by the formal or mainstream economy in the process of modernisation. The current use of flexible forms of employment and production, even in high-tech industries, now poses serious doubts about such an assumption. Certain types of informal or casual work are now an integral part of the modernisation process itself: they arise precisely because the introduction of new technology makes decentralisation of work commercially viable and efficient.

Flexibilities in employment in working hours or in locations of work could be in accord with the interests of workers. However, for socially vulnerable groups, flexibility often results in the casualisation of employment and in unfavourable conditions of work. Indeed, it is not only the changing structure of production but also the relationships of power at work and at home that affect the process of casualisation. The overall subordination of women in the society contributes to women's vulnerability as workers. Even when an industrial sector is organised and regulated, women in it tend to be offered insecure and casualised jobs. Cultural attitudes combine with social hierarchies; women are often employed because they are regarded as pliable, docile and likely to accept the negative consequences of flexible work. These social and cultural pressures can be exacerbated by political processes. State policies, at times, contribute to the process of casualisation. Workers in the Free Trade Zones, for example, are often not protected by a country's normal labour and employment legislation. They thus experience an informalisation of employment contracts even as employees of multinational corporations.

Informalisation also occurs when women, in the absence of job opportunities in the formal sector, take up self-employment as a measure of survival. These self-employed are not the small business people of the official economy, who can boast of being their own boss. They are the petty traders who have no control over their market; who can sell their services and wares only if they comply with the terms dictated by the subcontractors of large retailers. In the absence of adequate collaterals, they are invariably excluded from the credit facilities of financial institutions. By providing credit at an extortionate rate of interest, money lenders and suppliers of raw materials at times reduce the so-called self-employed to the status virtually of bonded labourers.

The informal sector is not a peculiarity of the Third World; neither is it a sector totally dissociated from the formal one. The growing casualisation in the developed countries in the shape of part-time and temporary work has also given rise to a dramatic increase in the types of employment that represent enclaves of informality in the midst of the formal sector. There are degrees of informality, depending on the extent of casualisation in the conditions of work.

Mirroring the developing world: flexible working contracts in the developed world

In order to fulfil her role as a mother and a homemaker, a woman often accepts vulnerability in order to achieve flexibility in working hours. It is in search of this flexibility that women in the richer part of the world undertake part-time employment and temporary jobs. Yet these are the very jobs which confer limited security to recruited workers. Since the 1980s, the majority of new jobs in the west have been of a casualised nature.[5] The trend towards casualisation has been intensified by the current changes in the organisation of work. Much of the employment and labour legislation has been viewed by employers, both in the private and in the public sector, as a fetter to the creation of employment. Hence, governments of most Organisation for Economic Co-operation and Development (OECD) countries have actively encouraged deregulation as well as privatisation of the social means of production. One consequence of such a trend has been an increased reliance on subcontractors by private as well as public corporate organisations. For management, the policy proves cost-effective. Workers in the smaller subcontracting units are entitled to far fewer benefits and legal rights than their counterparts in big organisations. The spread of subcontracting receives added impetus also from a changed management strategy, that tries to minimise the labour market inflexibilities arising out of union militancy and increased wage demands.

Italy is often cited as a successful model of this emerging decentralised mode of production. The pattern of work organisation as observed in Italy is quoted, in management literature, as flexible specialisation, achieved through an efficient subcontracting of work by big corporate organisations.

Italian employers, in response to the strike waves of the 1960s and 1970s, resolved to redress the 'Italian' mistake of relying on unionised workers of large corporations. The result was the formation of networks of large companies and their small subcontractors. These networks proved a cost-effective way of organising work and contributed to Italy's impressive growth in manufacturing. In 1987, the per capita income of Italy – once described as the sick man of Europe – surpassed that of the UK. 'Il Sorpasso' became the proud message of Italy, tempting other European and newly industrialising countries to follow its proven path.

Italy's example is often defined as the 'Benetton model'. Benetton, the well-known clothes manufacturing company of Italy, represents the quintessential success of flexible specialisation.[6] The Italian experiment with the decentralisation of work is close to the Japanese Just-in-Time (JIT) management strategy. The secret of success in Japanese industry has been its ability to have materials and manpower 'just-in-time' rather than 'just-in-case'. The stockless production, Japanese style, is dependent on an efficient network of local subcontractors that can be relied upon for timely and speedy delivery of materials, on a daily basis, to the main producer. That Toyota depends on 36,000 subcontractors for its supply of materials has assumed the proportion of folklore in the western world. The trend has been to follow Japan's example of 'stockless production', in the developed as well as in the developing countries.

The 'Just-in-Time' system is being facilitated by current trends in technology. Advances in technology make it easy to manufacture in small production units even those electrical and metal manufacturing goods which, until recently, needed factory-based production.

Flexibility in the organisation of work has become a key business principle in the contemporary, so-called post-Fordist era, where diversity and 'economies of scope' are viewed as more important than price competitiveness and 'economies of scale' as strategies for survival.[7]

A flexible response to market signals is being achieved through technological innovations, in the shape of computer-aided designing and manufacturing systems. This flexibility of small-batch production is achieved also through organisational innovations of decentralised production.

In some cases, large enterprises cease to produce entirely, subcontracting even the assembling and packaging aspects. The main companies, however, zealously retain their hold on marketing and on brand names. This centralisation of marketing, coupled with decentralisation of production, gives rise to what is known as hollow corporations.

Numerically, women have gained from these trends towards flexibility as they have been the main recipients of various forms of casualised jobs. Nonetheless, it is the deteriorating terms in the contracts of employment that have alerted women activists, particularly in trade unions: 'For women, flexible working often means greater insecurity, reduction in working hours and pay, changes in shifts, loss of national insurance benefit, loss of overtime bonuses, and loss of holidays, maternity leave, sick pay and pension.'[8]

Flexible specialisation in developing countries

The Italian or Benetton model nonetheless has already captured the imagination of some developing countries. In their attempts to become competitive in the world market, the government of the Republic of Cyprus, for example, has stressed: 'We are thinking in terms of the Italian model rather than the Korean and the Taiwanese. That means flexible specialisation where you cater for a high quality market like Europe: the Benetton (the Italian clothing company) approach.'[9]

With or without government support, an incipient flexible specialisation is already in operation in many parts of the developing world. The garment and textile industry of Aguascalientes in Mexico is one such example. Here, large companies, involved in a *maquila* (subcontracting) operation, maintain firm control over design, packaging, labelling and quality control – those aspects of production which are essential for maintaining brand names. They also keep firm control over the market. It is only the subprocesses of the garments that are farmed out to smaller subcontractors and outworkers. Small subcontractors in their turn seek labour from the outworkers in busy seasons or when they need a specialised service, such as adding an embroidered design to a garment.

The garment industry of India, likewise, follows its own brand of flexible specialisation. The industry employs a limited number of workers in the organised sector. They work in a modern factory environment with the latest technology. They are mainly skilled, well-paid cutters or tailors. These workers are always men, but the majority of the labour force is made up of unorganised, unskilled, women and men machinists belonging to the Muslim community, who stitch parts of garments at home.[10]

A separation between skilled and unskilled operations is crucial in following an effective subcontracting strategy. With certain products, where the manufacturing process is simple, the product does not undergo much transformation at the main production unit except for labelling and packaging. But for goods which have a more complex manufacturing process, skilled operations are often done by permanent 'core' workers in the main factory premises to ensure quality. Thus, the lace makers in India are predominantly homeworkers, but the finishing and stretching of the lace articles are always done in the houses of the traders. These skilled jobs are done by permanent workers who enjoy the confidence of the exporters/traders.[11]

The organisational innovation of stockless production has become a necessity in many countries, such as Brazil or the Philippines, in order to remain competitive in the export market so that they can pay back their foreign debt. The trends towards ancillarisation in the case of the automobile industry in Brazil[12] or in the garments industry in the Philippines[13] are well documented. For the developing countries, the pressure to reduce costs by streamlining the production process becomes increasingly urgent as computer technology gradually erodes their comparative advantages based on the cheapness of their surplus labour.

In India, there has been a steady trend for smaller production units to become ancillaries of large companies. The trend is particularly visible in light consumer goods manufacturing, a sector that provides much of the employment for women outworkers. Large and small units previously shared 'common markets' on the basis of the sector's horizontal division; they now share a 'common production process' as different stages of production become fragmented and specialised. This is what economists define as a vertical integration of the production process. In Calcutta, India, for example, small units for a long time produced ready-to-sell electric fans; more and more of those units have now switched over to working on orders for small parts or for specific production processes of large and well-known producing firms. As Nirmala Banerjee in Calcutta observed:

> This means that (for small ancillaries) the problem is no longer that of finding raw materials at reasonable prices . . . or of marketing their final products. They are now more concerned with getting adequate orders by undercutting other competing small units.[14]

The consumer electronics and automobile industries also strive towards a JIT system in many developing countries. The production of spare parts and components accounts for an average of 60 per cent of employment in the automobile and the consumer electronics industries in India. However, these items come mainly from the subcontracting units. Not all, but many of them operate in the unorganised sector, where the wages are low and benefits are nil.[15]

The changing pattern of work organisation affects the gender structure of employment in the developing world. A survey of the Coimbator district in India, which has taken to powerlooms in a big way, shows that an increasing number of powerloom weavers are now working on orders from the composite textile mills of the region, which could increase production without paying wages at mill rates. In these small non-unionised production units, the rate of increase of employment for women has been impressive. This is in sharp contrast with the employment structure in the large mills, where women workers represent no more than 4 per cent of the total workers and where their share is decreasing all the time.[16]

The greater the levels of subcontracting, the smaller is the unit of production at its further links. Women concentrate mostly at the weakest links in the chain of subcontracting. The clothing, footwear and shrimp-processing industries in Mexico and India are much-cited and good examples of such a phenomenon, where homeworkers ultimately provide the bulk of the labour inputs.[17]

Since the mainstream union movement has not, as yet, taken the issue of casualised workers on to its agenda, women can rarely look towards unions for improving their bargaining power. As a homeworker in the footwear industry in Calcutta states: 'Unions cannot fight to improve our lot. Their brief is to bargain for the permanent employees of the large-scale organisations.'[18]

Internationalisation of hollow corporations and women workers

The phenomenon of hollow corporations is visible most conspicuously in the export-oriented industries of the developing world. These include industries like garments and footwear; they also cover handicraft industries which cater for the refined taste of a certain section of the western public, weary of mass-produced, standardised products. Both categories of industries rely heavily on workers of the unregulated sector.

The inclusion of casual workers in the chain of subcontracting relates to a specific way of organising business efficiently for the world market. In the 1970s, for the transnational companies (TNCs) the move was to locate labour-intensive parts of electronics, footwear and garment production in low-wage countries that had a virtually inexhaustible source of cheap female labour.

In the pursuit of cost competitiveness, foreign companies set up branches or subsidiaries in the developing world; they also went into joint ventures with local businessmen. This proved a convenient arrangement for the transnational companies as they became by far the more powerful partners, by virtue of their access to world markets and technology. This specific mode of international sourcing of goods perceptibly changed in the 1980s. A system of networking with local subcontractors became the focal point of the new pattern in international sourcing; an emphasis was given to minimising the production role of TNCs as much as possible. Thus, it was the local firms that became solely or mainly responsible for production; but they remained dependent on large foreign companies for markets, materials and technical know-how. Local firms simply performed the ancillary roles of processing orders and/or materials supplied by the international corporate firms.

The new mode of sourcing became attractive to transnational companies for two major reasons. To start with, international subcontracting reduced the visibility of the TNCs in terms of direct equity investments; thus it avoided the problem of nationalisation as supplies came mainly from firms wholly owned and operated by the locals. TNCs could thereby move away from equity control and yet could keep stricter reins on market and technology. The second major advantage of relying on numerous local subcontractors was that it gave the international companies the ability to respond flexibly to changes in market demand. This characteristic became an attractive attribute in the business world where, in certain sectors, ability to cater to changes in fashion and design became an extremely important strategy for survival.

An increased reliance on local small-scale subcontractors diminished the significance of export-processing zones and large factories as sources of employment for women. Their declining importance is visible in the changing pattern of direct foreign investment by the market economies of the developed world. In 1975, 42 per cent of such investment went to developing countries; their share had fallen to 17 per cent by 1986;[19] this deceleration of investment still continues.

To a certain extent, technological changes explain this phenomenon. With rapid introduction of computer-aided technology, differences in labour costs in different countries have become less important in determining investment by multinational companies, but increased publicity for unfair working conditions and resistance to unionisation in the export-processing zones has also made the companies sensitive about their image in the wider world. The workers employed by the local subcontractors remain flexible and invisible. In contrast, women of the export-processing zones are increasingly getting organised and establishing links with labour movements outside the country. The protest in 1989 against the refusal by a British company to pay a legally stipulated minimum wage to the workers of the Bataan export-processing zones in the Philippines is not atypical. The international adverse publicity led to the company's closing the factory in the zone, making nearly 1,000 women unemployed. As Nancy, one of the Bataan workers on the picket line protesting the closure, states: 'The Government's export processing zone authority says our camping and banners might deter investors. But the very reason we are on the picket line is to keep the factory open.'[20]

Hollow corporations and casualised workers

The dependency relationship between the minuscule local subcontractors and the giant international or national retailers characterises the present mode of export-oriented manufacturing and handicraft production. The garment industry is the most conspicuous example of the effect of international subcontracting on the quality and quantity of employment in the developing countries.

Export-led production significantly improves the employment opportunities of women in factories as well as in home-based work. To that extent, international subcontracting alleviates the unemployment problems that generally beset the poorer world. The governments of developing countries, understandably, view their countries' link with the international subcontractors as a cost-effective way of creating additional employment. Hence, the local exporters and subcontractors receive active governmental support and finance.

In spite of its beneficial effects on the level of employment, there has been a growing concern over the phenomenon of international subcontracting. To start with, the local subcontractors are themselves often powerless in relation to the exporter or to the international retailers.

The subcontracting system depends essentially on price bidding. Since there are many competitors, the dealers bid the lowest possible price in order to get the contract. The result is that, after meeting his own overhead costs, the subcontractor is often left with inadequate money to pay out decent wages to piece-rate workers.

Subcontractors generally retain strict control over the working patterns of casual workers both through economic and extra-economic channels. In Rio de Janeiro, the intermediaries or subcontractors ensure the return of material advanced to homeworkers through retention of voter registration cards.[21] Intermediaries also utilise their personal contacts in the homeworkers' community to exert control and supervision. The contact person in the community thus becomes responsible for the assembly of materials within a designated time period, and according to factory specifications. This type of networking stimulates the creation of a long chain of jobbers, contractors or subcontractors acting as intermediaries between sweatshop workers, homeworkers and the final employers.

Women's survival strategies and the growth of casualised work

The spread of casualised work often arises from poor women's desperate need to survive outside the organised sector. In spite of its bleak aspects, it is informal employment that has acted in the last decade as an important shock absorber against the economic crises of the developing world. As the traumas of the debt crisis and structural adjustments hit the stability of the official economy, workers are forced to find their livelihood increasingly in the unofficial one. This trend is visible especially in Africa and in Latin America.

To pay for the dream of rapid modernisation and urbanisation, the developing world has gone into debt since the mid-1970s on the massive scale of one trillion dollars. The interest charged by the northern banking system means that the developing world now pays back more than it receives in aid and new loans combined. In order to meet payments of this order, in many countries land for food has become land for growing export crops. The liberalisation of external and internal trade has meant the collapse of many a domestic industry and the withdrawal of food subsidies. The consequent structural adjustments have led, therefore, to unprecedented rates of inflation and deindustrialisation (see Figure 1.1 and Table 1.1).

In Africa, the effect of the debt crisis has been compounded with a fall in the prices of all major export crops (Figure 1.2). In this deteriorating situation, women and children pay the highest price in malnutrition and vulnerable informal employment (Table 1.2).

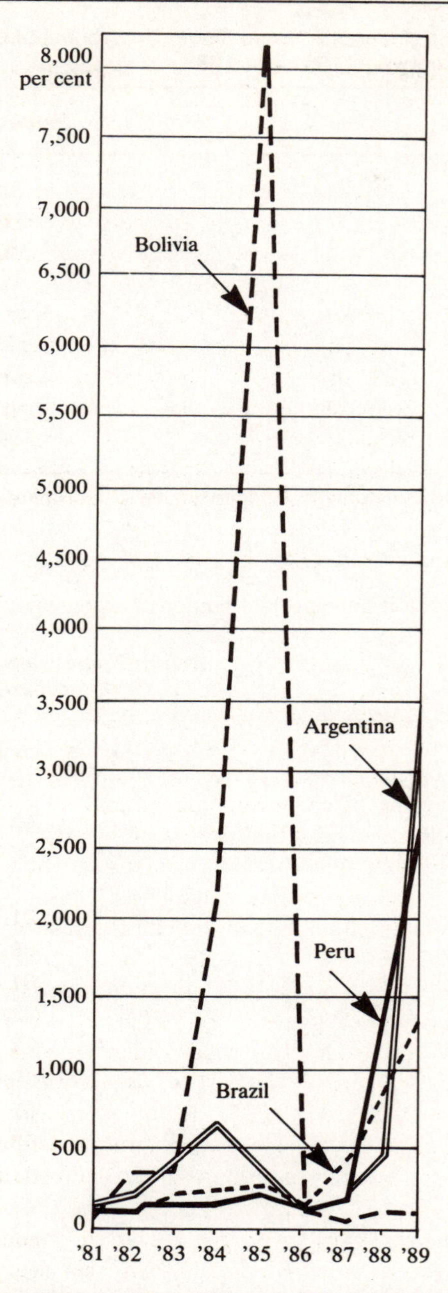

Figure 1.1 A dismal decade: hyper-inflation of Latin American countries

Source: International Financial Statistical Yearbook, IMF, Washington, 1990

Table 1.1 Changes in per capita Gross Domestic Product of Latin American countries between 1981 and 1989

Country	*Percentage*
Mexico	–9.2
Venezuela	–24.9
Columbia	13.9
Brazil	–0.4
Ecuador	–1.1
Peru	–24.7
Bolivia	–26.6
Chile	9.6
Argentina	–23.5

Source: Statistics given by Economic Commission for Latin America and the Caribbean, Santiago, Chile, 1990

Table 1.2 Informal sector: an important source of employment: Africa

Country	*Informal sector employment as % of urban total*
Benin	72.6
Burkina Faso	60.2
Burundi	45.1
Congo	36.9
Côte-d'Ivoire	60.8
Gabon	21.8
Ghana	38.3
Guinea	61.2
Madagascar	22.7
Malawi	23.0
Mali	32.9
Nigeria	65.1
Niger	68.5
Rwanda	47.2
Senegal	44.3
Togo	60.4
Zaire	66.2

Source: ILO/JASPA, African Employment Report, Addis Ababa, Ethiopia 1988

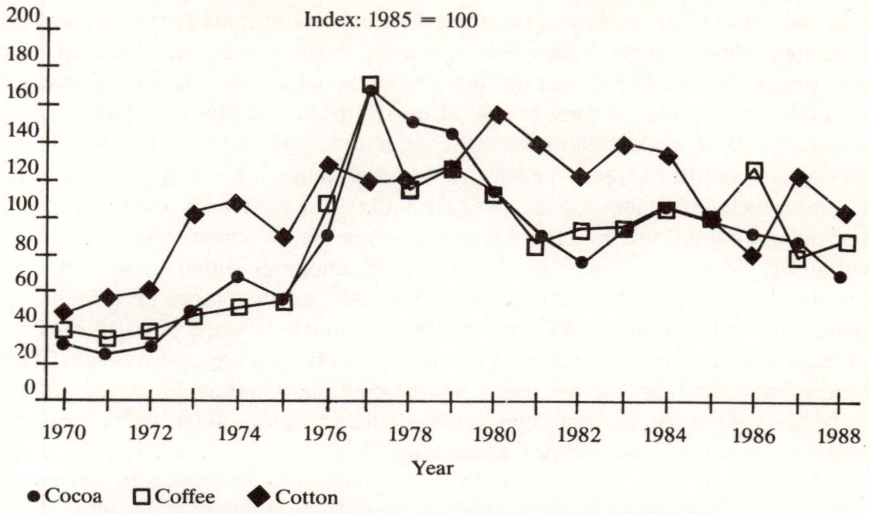

Figure 1.2 Export prices for three major sub-Saharan commodities

Source: International Financial Statistical Yearbook, IMF, Washington, 1990

In the face of such crises, people at the grassroots level do not wait idly for help from the world outside. They find ways of defending themselves and their families against acute tribulation. The reliance on one's own resources gets exacerbated as the ordinary people face the mercantile state which, to quote Mario Vargas Llosa, the Peruvian writer and politician, 'legislates and regulates in favour of small pressure groups and discriminates against the interests of the large majority'.[22] In Argentina, the unreported informal economy in 1988 produced $50.4 billion, as against the official Gross National Product of some $70 billion. In Kenya, the unregistered minibuses, called *matatus*, dominate the transport system. In Peru, according to Hernando de Soto, the champion of the Peruvian unorganised sector, the informal economy adds 29 per cent or more to the official Gross National Product.[23]

Women (and men), being displaced from the official economy, often set up on their own with a little capital and a lot of initiative. In Tanzania, for example, as Marja-Liisa Swantz recounts,

> they start selling small items: soap, salt, trinkets, soft drinks. Eventually the small business grows into a *genge*, a little stall with increased varieties of goods. Some bleach brown salt and grind it, others make ice cream and 'bites' of all kinds, from rice cakes to donuts and meat pastries . . . the inventiveness also increases with the need . . . the mother sends her little sons to collect inner tubes and cuts from them elastic bands for the underwear she sews and sells.[24]

A spirit of co-operation, solidarity as well as struggle permeates the initiatives of this type in the unregulated sector, providing the seed of a new type of grassroots mobilisation.

In south and south-east Asia, the green revolution contributed to casualisation of women's work. The few new jobs that were created in modernised farming went primarily to men; it was the male workers who were trained to handle mechanised operations, the use of chemicals and the newer varieties of seed. The changed mode of production meant that even traditionally feminised jobs such as dairying and poultry keeping became male occupations. Thus dispossessed, the unemployed female labourers desperately looked for paid work to supplement their meagre family income. For the employers in the manufacturing sector, the availability of a new source of cheap female labour provided them with an opportunity to change from factory-based production to a more decentralised method of work organisation. For displaced women labourers too, the subcontracting arrangements opened up a new possibility of wage employment. The work, more often than not, proved to be low-paid, hard and exploitative; yet as the homeworkers and the employees of the subcontracting units themselves so emphatically said, 'it was better than nothing'.[25]

Women and manufacturing homework

Homeworking, as a form of wage employment, highlights the dilemmas that women face in the sphere of paid work. Even in the small factories of the unregulated sector, it proves difficult to combine paid work with the need to look after the very young:

> Fatima works in a small shoe factory in Calcutta. She runs a mile every day to reach home in order to breastfeed her infant during her lunch hour. Her unemployed husband Rahim suffers from dengue fever, which at times leaves him unconscious. It is not easy for him to bring the infant to her place of work. By the time Fatima finishes feeding the baby and runs back to her shoe factory, she is invariably late. The manager scrupulously notes the time lost and deducts money from her weekly wage packet. Fatima now seriously considers homeworking – it will help her to look after the baby while at work.[26]

This flexibility, however, has a price. Homeworkers are usually piece-rate workers. In order to earn a living wage, a homeworker has to drive herself hard. One knitter in Uruguay succinctly articulates what she observes to be the main problems of homework:

> When you work outside [the home] you comply with the work schedule there and then you come home . . . but here, with this work, there are no holidays, nothing . . . because every day you have to do it . . . you do not rest Saturdays and Sundays as in other jobs . . . sometimes Sunday is the worst day because we have to hand in work.[27]

The pace of work, even at home, is determined by the contracting agent or the middleman. Constrained by deadlines imposed by the ordering firms, the subcontractors have to ask the homeworkers to meet the quota at any cost. During periods of heavy demand and tight production schedules, homeworkers report going without sleep, with the work day extending from ten to twenty-four hours a day. In order to finish in time, children participate in the work effort, accounting for a large pool of unpaid family labour. When there is inadequate bargaining power, women are forced to relegate the unpaid domestic work to the dictates of paid production. Any demonstration of militancy or non-cooperation may mean an end to the supply of work. With a large pool of poor women who are desperate for a livelihood, it is easy for the subcontractors to find homeworkers elsewhere.

The ease with which homeworking can be shifted to another region, or to different sets of workers within the same region, makes employees wary of demanding legal rights, even if there are some. In Sri Lanka, where home-based *bidi* (small cigar) workers are entitled to certain legal protection, it is they themselves who often collude with employers to negate the provisions. In the face of intense competition for jobs, workers agree not to assert their rights in order to remain in the good books of the employers.[28]

The clandestine economy and the informal sector

The most disturbing aspect of casualised work is the way it can get enmeshed with the clandestine economy, where workers have little redress against either physical or economic exploitation. In both the developed and the developing parts of the world, there has been a steady increase in the size of the underground economy throughout the 1980s. A widespread concomitant increase of homeworking, particularly in the manufacturing sector, has been its predictable outcome.

In the richer part of the world, it is the immigrants from the developing countries who get readily recruited into this form of clandestine work. In the United States and Canada, the jewellery and garments industries are well-known for their use of undocumented industrial homeworkers from the Third World. The industry is seasonal, with fluctuations in production and employment. The contracting systems, in this situation, encourage the undisclosed employment of industrial homeworkers. As was the case in the nineteenth and earlier twentieth centuries, subcontractors now underbid one another by hiring employees willing to work for less money. Today, those employees are overwhelmingly the poor immigrant women of Toronto, New York, Chicago and Los Angeles. Often, they are undocumented workers who come from South or Central America; many are Asians: Chinese and, increasingly, Korean. This undocumented workforce forms an underclass that provides the cheapest of labour. The exact number of such homeworkers is difficult to estimate, as the employers operate illegally on the fringes of the economy, in order to avoid unemployment insurance, minimum wage rates and regulations concerning child labour.[29] The average wage can be 80 cents or less per hour. For an undocumented worker it can be difficult to demand and receive even that. The veiled threats of deportation for workers who do not possess 'green cards' prevent them from making trouble even when they are not paid their due wages.

The use of migrant workers as illegal homeworkers is equally common in western Europe. 'Guest workers' from Turkey or the former Yugoslavia who have outstayed their work contract are typical employees of the hidden economy of Germany and Holland. The employers play on the fear of the employees to get work done at a rate that is often cheaper than what prevails in many so-called Third World countries. Holland, once a flourishing centre of textiles and garments production, now hardly possesses any large-scale factory in this specific sector. In place of formal factories, Amsterdam has experienced an unprecedented rise in hidden homeworking and ethnic sweatshops involved in garment production.[30]

In east and north London in the United Kingdom alone, there were about 20,000 immigrant women working in the hidden sector of the clothing industry by the mid-1980s,[31] and this process has continued. Some take up homeworking as it is culturally acceptable, particularly in the Muslim communities, but most of them would prefer to have jobs outside, if it were not for the uncertainty of their status as immigrants and for the fear of racist attacks.[32]

In the developing world, the steady rise of homeworking has been similarly intertwined with the spread of casualised employment in the clandestine sector. According to one estimate, in 1985 in the garments industry alone, 22,000 employees were working underground in Buenos Aires and 78 per cent of them were women. The majority of them were working from units employing no more than two to five people.[33] This increase in undisclosed work has not been the result either of an increase in production or of a rise in the levels of subcontracting. From the mid-1980s, a large proportion of the subcontracted work has simply gone underground. The experience of Argentina exemplifies that the most effective way of avoiding the costs of redundancy and social security payments, on the part of the employers, has been concealed employment, either within the firm or, better still, at home-based units.[34] This spread of an underground economy has also been noted in other parts of Latin America, such as Mexico, Brazil and Peru. In Colombia, a very rough estimate points to a steady increase in non-protected employment from 13 per cent in 1975 to 35 per cent in 1980.[35] Women, as always, are the major recipients of such unprotected jobs, which include concealed home-based work.

The garments industry in most Asian countries likewise thrives on the hidden labour of the underground economy. In Bangkok alone, there are about 1,500 to 2,000 minuscule illegal factories, producing for the export-oriented ready-to-wear garments industry.[36] In Delhi, India, only 13,500 workers were officially reported to be employed in the garments industry in 1983. Research undertaken by Rukmini Rao and Sabha Husain shows, however, that the true figure could be well over 100,000. Most of the workers are 'hidden' employees and a large number of them are home-based women.[37]

Child workers are commonly found in home-based units and small firms. In Latin America, the production of garments and furniture in the unregulated sector relies heavily on the labour of child workers. In Mexico, children as young as 6 engage in the assembling of toys, plastic flowers, cardboard boxes and metal-ware. They are mostly girls, who work either as homeworkers themselves or as assistants to their homeworking mothers.[38]

The bulk of the child workers in the Philippines are domestic outworkers. They definitely form a regular reserve labour force for export-oriented garment production.[39] Girl children are socialised into stitching from the age of 2 and are fully absorbed into the workforce by the time they are 4. Besides their lost childhood, young workers suffer also from various health hazards. The use of child labour in the factories and home-based units of the carpet, rug, leather tanning and gem-polishing industries is well known. The nimble fingers and dexterity of young children are highly prized in these industries – but these are not the only reasons for seeking their unorganised labour.

Poverty forces children to seek jobs in the undisclosed services sector. Young girls are put out to work or sold as servants, maids or prostitutes in many parts of the developing world for a pittance, working throughout the day and sometimes for a good part of the night. In the red light districts of Manila, capital of the Philippines, $10 buys sex with very young girls. A BBC 2 programme on child slaves (23 June 1989) documented the life story of two such average girl workers of the informal sector:

> Eva was only 10 when she went on the game [of prostitution]. She is now 13. Her friend Josie, now 10, started when she was 7. Their best customers are sailors from the Australian Navy. The girls, who have taken to sniffing glue to escape their reality in child prostitution, hate their work in a haven for paedophiles. But they need money to buy food and clothes. And not just for themselves – many of the older girls have children of their own. There is no other available work they can do.

The economic crisis of the 1980s and increased impoverishment in certain regions of the developing world have given rise to an alarming increase in the supply of child labour. Films such as *Pixote* and *Salaam Bombay*[40] show graphically the price that children pay in a country's attempt to gain rapid modernisation. In the poor part of the world, working children are visible on urban streets, engaged in loosely economic activities such as vending, cleaning shoes and washing cars, as well as prostitution, drug trafficking and other illicit activities. Many, aged between 7 and 17, end up essentially living on the streets. They are exposed to exploitation, violence and discrimination that seriously endanger their physical and psychological well-being. As a group, they are commonly known as street children; they are the most endangered working children of the urban areas.

Beyond the domain of legal economy

In discussions among economists and political philosophers, the concept of the informal sector has in recent times assumed an important ideological dimension.

Some have glorified the role that it can play, stressing the entrepreneurial potential of the workers themselves in this sector. Yet the workers, often described as mini-entrepreneurs, are, in fact, desperate workers, bereft of a job as well as of capital. Improving the future of such a workforce calls for measures beyond the mere debureaucratisation of the state machinery or the formation of self-help groups.

When employed in jobs outside the periphery of the legal domain, casualised workers, being accomplices in an illegal act, face an added vulnerability, not having access even to the rudimentary civil rights that the state offers. For a worker in the illegal sector, it is impossible, for example, to seek redress against physical harassment meted out by the immediate employer.

This erosion of basic civil rights is most noticeable now in certain parts of Latin America and Africa, which are experiencing the painful traumas of debt crisis and structural adjustments. In these regions, the illegal informal sector has become the growth area of economic activities accounting for an ever-increasing share of the countries' employment. In the view of some thinkers, this development does not necessarily herald a bleak future. To Hernando de Soto and Mario Vargas Llosa of Peru, for instance, the new trend indicates a silent revolution. They have hailed the spread of the informal sector as the Other Path (*El Otro Sendero*), an alternative to the bloody revolution that movements such as the Shining Path (*Sendero Luminoso*) stand for. They maintain that the silent revolution represents people's protest and provides a countervailing power to corrupt bureaucratic and centralised state power. Hernando de Soto writes, in *The Other Path*, of:

> the elephant-sized (Peruvian) bureaucracy that, in order to justify its existence, requires, for example, that to register a small company a citizen must contend with eleven different ministries and municipal offices over a period of ten months. . . . Laws are cooked up in the bureaucratic kitchens of the ministries . . . in accord with the persuasive power of the interest they will serve. . . .
>
> Hence people have renounced legality and taken to the streets, selling whatever they could, setting up their little shops and building their own homes on the hills and sand dunes. Where there was no work they created it. They made a virtue of their shortcomings, and turned ignorance into wisdom.[41]

As de Soto proclaims, 'We [the supporters of the informal sector] have become a movement.'

This optimism, however, needs close and sharp scrutiny. Even leaving aside the criminal activities such as drug trafficking and related trades, any illegality in the informal sector hardly bodes well for the average worker. In the absence of any organisational strength, the workers understandably become subject to the manipulation of those who control the market and finance in this parallel sector. The contribution of the illegal informal economy to the Gross National Product of some Latin American and African countries may be impressively high, but so are the numbers of child workers and of employees in precarious jobs.

It is misleading to declare, as de Soto does, that the informal sector represents the initiatives of people using illegal means to achieve legal objectives. People do not form an undifferentiated mass or homogeneous group, sharing the same economic objectives and similar economic power. They are in an unequal relationship of power as a result of the undemocratic distribution of the means of production and unequal access to social and economic resources. This is precisely the reason why women workers need organisational strength, to achieve security of employment as well as protection from work-related hazards – in the formal as well as in the informal sector.

ON ORGANISING CASUALISED WORKERS

The need for mobilising

Workers of the unregulated sector are often justifiably called invisible workers. When they work in isolation in sweatshops or in scattered home-based units, they receive neither the benefits of employment legislation nor protection against industrial health hazards. In the absence of a collective identity, it proves difficult, if not impossible, for the workers of the unorganised sector to receive much protection, even when the laws are in their favour.

There is no dearth of legislation, for example, that aims to protect homeworkers in different parts of the world. In the context of a select number of countries, *Conditions of Work Digest on Homework* (1990), published by the ILO, gives a summary of the relevant legislation. Admittedly, there are variations in the degrees of protection offered by different countries. In Peru, for example, the law stipulates that the wage rate for homeworkers cannot be less than that paid for the same work in the factory or workshop. In the Dominican Republic, in contrast, a homeworker's wage is determined by agreement on rates payable for the task. In India, only factory workers are subject to the State Insurance Act; homeworkers are excluded from this and from the laws regulating contract labour. There is thus a need for upward harmonisation of protective laws in most countries. But improved legislation can provide only an inadequate redress unless and until workers' organised actions help to enforce it.

The experience of Italy is illuminating in this context. In Italy, since 1973, homeworkers have legally been able to demand the same benefits and wages as those enjoyed by the factory and office workers. In order to avoid the implicit financial costs, however, it has become customary for the employers to encourage home-based workers to register themselves as self-employed artisans. The result has been a suspicious increase in the number of artisanal firms over the years, as well as an unconvincingly high share of artisanal units in the total production. Matelda Fedi, a trade unionist from the Confederazione Italiana Sindacati Lavoratori (CISL), Tuscany, Italy, sums up the situation clearly: 'for Italian unions, the issue at present is not to demand more legal rights but to ensure that they can be applied'.[42] There are two main reasons why it is often difficult to enforce the progressive laws. First, even in the richer countries, lack of resources besets the machineries responsible for the enforcement of the enacted laws. Thus, in the United States, the understaffed inspection force finds it difficult to check for the presence of illegal homeworkers and child labourers.[43] Second, opportunistic employers can always find ways of evading the existing law. With professional advice, is is not difficult to identify loopholes in any piece of legislation. By making use of them, employers can stay on the right side of the law, while consciously infringing the spirit of the law. In addition, by transferring work into the black economy, it is possible to evade most regulations.

Co-operatives based on workers' participation

The difficulties encountered in enforcing legislation have encouraged the policy makers in several countries to assist in the formation of workers' groups. The help comes primarily in the form of encouraging self-employed poor women or homeworkers to start co-operatives or self-help organisations.

Significantly, the formation of participatory co-operatives has so far been successful only in the areas where the trade union movement had already been strong. The Dinesh *bidi* co-operatives of the Cannore district of Kerala in India exemplify this crucial role of unions. The co-operatives started in 1968, when the private commercial entrepreneurs left the district in response to the 1966 Bidi and Cigar Workers' Act. The Act gave *bidi* homeworkers employment rights on a par with factory workers. The cost of hiring homeworkers rose, and private employees left the business, heralding unemployment for 12,000 home-based workers. The state government responded to that situation by organising workers in a series of producers' co-operatives and giving loans to workers to buy shares and raw materials. The co-operatives started with 3,000 members in 1968; by 1983, the membership had grown to 30,000. In the co-operatives, workers received fair wages, maternity leave, group insurance and retirement benefits. All in all, they proved an immense success and were viewed as worthy of replication in several parts of India.[44]

The replication, initiated by governments, however, proved far more difficult. Besides the local characteristics of the workers and the market, the secret of success of the Dinesh *bidi* co-operatives lay in the strong trade union movement of Kerala. The strict monitoring of the implementation of the Bidi and Cigar Workers' Act by unions encouraged the private manufacturers to desert the area, leaving the market entirely to the workers. In the non-unionised regions, in contrast, the private sector still functioned, relying on clandestine labour; and thereby undercutting the co-operatives who paid fair wages and the central excise tax.

The comparative experience of Kerala and other regions has thus led some researchers to conclude that a pre-existent 'organisation of workers is a necessary precondition for the formation of workers' co-operatives in the developing world'.[45]

The main task, therefore, is to devise ways of mobilising the workers of the unorganised sector through structured trade union organisations. Even if at times this challenge seems insuperable, some amount of success in this regard in south Asia has offered hopes in this direction. The two best-known organisations of this nature are Working Women's Forum (WWF) and the Self-Employed Women's Association (SEWA), both in India. They are both registered as independent trade unions, and over the last decade have recruited nearly 85,000 and 46,000 workers respectively. Besides home-based workers, their membership consists also of vendors, petty traders and hawkers or labourers selling their services in plantations, catering, laundering and similar occupations.

The programmes of SEWA and WWF are specifically targeted towards women's demands. They are based on an integrated development plan for women that includes child care, leadership training and forming women's own savings banks. Besides, they promote informal producers' co-operatives, in order to give greater strength and bargaining power to their economically vulnerable members.

SEWA has been particularly active in bringing the plight of casualised workers to the notice of the national and international policy makers. The organisation, therefore, has become a model to study for activists and trade unionists in all parts of the world. SEWA started in Ahmedabad, India, in the year 1972, when a group of home-based and petty self-employed workers came together to form a workers' association. It started as a women's wing of the Textile Labour Association (TLA) and functioned within it until 1 May 1981, when it was expelled from the TLA for taking up the cause of untouchable (*harijan*) women. It is now a registered trade union in its own right and organises struggles for improved work conditions, and fights against mass unemployment. But central to the empowerment programme of SEWA is the development of co-operatives in order to devise alternative production systems, credit facilities and better bargaining power for its economically vulnerable members. Their mobilising and organising strategies are depicted schematically in Figure 1.3.

The examples of SEWA and WWF are important in that they highlight the potential for forming powerful organisations of women workers when the right strategies are devised. The membership roll is impressive both for SEWA and for WWF. Yet it would be inappropriate to evaluate their achievement merely by the number of members recruited. Their achievement lies also in making the hidden workers of the industrial sector visible to the national and international policy makers.

It is difficult to predict to what extent these two relatively successful organisations can be replicated elsewhere. In order to use these organisations as models, the governments and trade unions need to evaluate their history carefully in the context of the cultural specificities of these organisations, the role of the charismatic leaders, and their dependence on aid from external and internal donor agencies.

Both these organisations are immensely fundable; hence, they receive grants and assistance from donor agencies from all over the world. It is highly unlikely that other grassroots organisations, even with similar potential, will receive similar funds. For the activists and policy makers, therefore, a survey of the success and failure of smaller or less well-known grassroots organisations will be equally important.

1 To ensure visibility –
 a) by organising women into units/co-operatives
 b) by giving publicity about their existence

2 To wage a struggle – for better conditions of work and pay

3 To involve women in development activities – by giving experience in banking, marketing and retailing

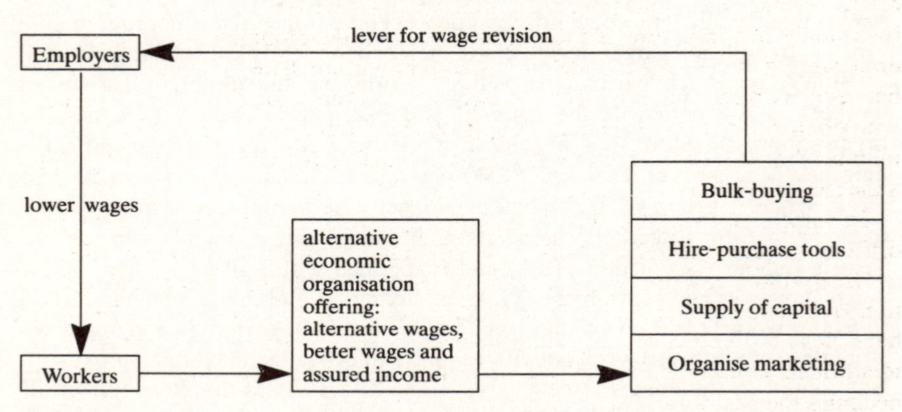

Figure 1.3 Strategies of the Self-Employed Women's Association in India
Source: SEWA (Self-Employed Women's Association), India

Co-operatives and the attainment of legal recognition for workers' associations

The benefit of being a member of a trade union-cum-co-operative organisation such as SEWA is that a casualised worker begins to perceive herself gradually as a proper 'worker', entitled to rights similar to those the state offers to the workers of the organised sector. In such a struggle for entitlement and for self-development, a close link between the trade union and the co-operative movement proves highly effective. To start with, both trade union and co-operative movements share the same goal, which is the improvement of the economic and living conditions of the working population. By joining together, the movements in many ways reinforce each other's ideals.

The benefits that these linked trade union-co-operative activities bestow on the workers are manifest in the improved bargaining power of casualised workers, as experienced by the women workers of SEWA:

> The contractor of the road construction gang was paying us only Rs 3 a day. We found out that the minimum wage we should be getting is Rs 11 or 7kg of grain. We decided to do *ekat* [solidarity]. We said we would not accept the

> wages till he paid us Rs 11 a day. He refused. We held out for two months, then saw the collector and told him of our problem. Ten days later the *mate* [contractor] came back and paid us the minimum wage and since then there have been no problems.[46]

A close link between the co-operative and trade union movements gives a workers' organisation a chance to achieve legal recognition without undue delay. A co-operative is defined as an institution which aims at social and economic betterment of its members through mutual aid, voluntary membership, a democratic system and equitable distribution. A co-operative therefore is an ideal institution for initiating income-generating, or developmental, projects for empowering the workers. Trade unions, in contrast, are there to regulate relations between workers and employers or between workers. For casualised workers, who often do not have one identifiable employer, hence are not easy to unionise, membership of a co-operative is often the most effective step towards achieving a measure of power.

In some countries, it is relatively easy to get registered as a union but more difficult to get registered as a co-operative. In India, for example, it may take up to twenty-three years to get a co-operative society registered,[47] but all one needs to be registered as a union is a membership roll of ten. In Malaysia, on the other hand, it is difficult to get registered as a union unless the members possess an identifiable employer. For the self-employed registering as a co-operative there becomes the best way of achieving a strong legal identity.

There are numerous informal co-operatives and unregistered workers' organisations in the developing world. Many groups, especially women's groups, often find that there are definite advantages in remaining unregistered, either as a co-operative or as a union. This way, they remain informal and flexible; and they do not face interference from government agencies. Moreover, self-employed, illiterate women feel more comfortable and more in control dealing with organisations where there are no formal procedures.

There are, however, overriding disadvantages in remaining an unregistered informal group.[48] It cannot, for example, acquire capital assets. Hence, if a private member wants to open a bank account, or an office, it has to be done in the name of that member. This places an extra burden of responsibility on the individual; it also opens up ways of private control and private gains. The lack of legal identity of an organisation also limits its access to many government schemes; in India, for example, many anti-poverty schemes launched by the government can in principle be used by home-based workers; however, the funds are obtained only if channelled through registered organisations.

The linked development of co-operatives and trade unions also helps the international donor agencies to channel funds to those income-generating projects which achieve, in addition to their immediate goals, a long-lasting solidarity and the confidence of the workers. The dignity of being recognised as a worker is important in the lives of home-based and other casual workers; it may be difficult to achieve this dignity in a project run by charitable organisations and informal associations.

It is not easy, however, to link women's co-operatives or self-help groups with the mainstream trade union movement. The difficulty lies in historical traditions. Until now, it has been customary for unions to engage essentially in wage negotiations that will secure a family wage for the male breadwinner. The very image of a male breadwinner has prevented the unions from taking up the issues that particularly affect the working lives of women. Because of a close relationship between their paid and unpaid work, women workers need to bargain for benefits which are wider than just a living wage. Yet, in male-dominated trade union movements, it can seem diversionary to take on such priorities and demands.

The visible success of some of the women's unions may mark a change in the future direction of union traditions. The *bidi* and tobacco workers' union in Nipani, Karnataka, India, in this context, deserves a special mention. Known as the Chikodi Taluka Kamgar-Mahasangh (CTKM), it is a home-based *bidi* workers union of a novel kind. It is basically a women's union with actions that go far beyond wage negotiation. Since its formation in 1980, its members have fought successfully for economic rights such as minimum wages and provident funds; they have fought equally powerfully on issues such as dowry, divorce and male alcoholism. Around the struggle against social and communal injustice, several supportive measures have been taken. A multipurpose co-operative has been set up to provide grains, kerosene and cooking oil to women at a cheap rate; there is now a small savings scheme to advance loans to women members at a reasonable rate, and a home for *devadasi* (temple prostitutes) and other women needing refuge.

The example of the Nipani *bidi* and tobacco workers' union shows that it may be difficult, but not impossible, to organise dispersed homeworkers. It also highlights the broader nature of the economic and social demands that bring women to unions.[49]

Traditional trade union techniques in organising the unorganised

The linking of co-operative and trade union movements, however, can impose excessive strains on the human, administrative and financial resources of unions, unless these are extremely well funded. Also, the strategy can be inappropriate for a vast number of workers who have identifiable employers, yet are not the beneficiaries of legal and employment protection because of the nature of their jobs.

Various other techniques for unionising unorganised workers can also be observed in recent initiatives in Latin America and Asia. In Mexico City, the Nineteenth of September garment workers' union has emerged, recruiting women wage-labourers-cum-housewives, employed in hundreds of hidden clothing sweatshops. The devastating earthquake of 1985 was the catalyst for the formation of this independent union: while the sweatshop owners in Mexico frantically tried to save their machinery and profits, there was little concern for the welfare or the survival of the labourers. Outraged at the employers' behaviour, the bereaved women got together to demand compensation for colleagues who died while at work. The women subsequently moved ahead to set up a trade union specifically for women workers in the clothing sweatshops. An invisible sector became visible overnight.[50]

The Nineteenth of September Union is the very first independent union to be officially recognised in Mexico. In demanding and receiving recognition, in 1986 the women organisers vigorously solicited support from the activists and committed academics of Mexico, Europe and North America (see Appendix I). Besides forming an international support group, the Nineteenth of September Union also emphasised the importance of changing the priorities in claims that are to be bargained for. Being a women's union, the provision of child care, for example, plays a prominent part in its agenda of demands.

For Mexican activists, the forming of this independent union has been only one way of organising workers. In some situations, women workers prefer to organise themselves in non-union institutions. At the Centro Obrero, Cecilia Rodriguez, the organiser of the Mexican garment workers in El Paso, Texas, explained that (often) women themselves become very anti-union because of their experience with yellow unions who worked in collusion with their employers and used violence to achieve their ends. She also said that one long-term aim was to build a strong membership organisation in which the members could build their own strategies, and to build up committees of five or six women in each factory to address the problem of that particular factory outside the mainstream union structure (see Appendix II).

In contrast, the unionisation of women homeworkers who roll *bidis* in the Indian city of Hyderabad has been remarkable. Thanks to the painstaking efforts and enormous tactical skills of women activists in Hyderabad, a trade union network among the *bidi* homeworkers in 1989 launched a successful all-out strike that forced the owners to accept the women's demands for increased wages and for identity cards as employees. As a homeworker there so optimistically states, 'we like the union; because of its help we can have two meals a day'.[51]

Innovative union strategies

Such new women's unions have proved effective in mobilising workers in forms of work which have been seen as well-nigh impossible to unionise. These approaches have a wide relevance. Alternative organising methods become necessary when workers lose their union membership because their jobs are put out to the non-unionised small-scale sector. As a report by the American Federation of Labor and Congress of Industrial Organizations (AFL-CIO) in the United States comments:

> These individuals might well be willing to affiliate with a union with which they have had contact or with which they have some legal relationship, provided that the costs are not prohibitive; this would be especially true to the

> extent that unions offered services or benefits outside the collective bargaining procedure.[52]

One innovative step could thus be a shift from enterprise-based unions, as in India, to universal professional or skill-based unions, as in Sweden, where, irrespective of the size and the location of the employer company, members can seek help from occupation-based unions. Such changes in the structure of recruitment do not happen rapidly. In the meantime, the developing countries may benefit by looking at the experiments in the developed world while reorienting their own mobilising efforts.

The Transport and General Workers Union, Britain's biggest trade union, aims to reverse the substantial fall in its membership by changing its image as a predominantly male organisation with little relevance to women. As a step towards this goal, the union has pledged to address the needs of the growing number of temporary, part-time casualised workers, the majority of whom happen to be women.[53] In a similar vein, the GMB in Britain has announced that in order to survive, the predominantly male British unions must put themselves forward as champions of the new 'servant' class of exploited female workers.[54]

The move towards self-criticism and changing strategies has been most successful in Canada. To bolster union organisation and make unions more attractive to unorganised workers, particularly women, the unions have actively pursued such issues as child care, pay equity and equal employment opportunities.[55] Unions in Canada have also taken affirmative action to increase the number of female representatives on their executives. The result has been encouraging: while in the United States the union density (membership as a proportion of the country's workforce) fell from 30 per cent in 1965 to a mere 17 per cent in 1989, it went up from 30 per cent to 35 per cent in Canada in the same period.

The growth of casualised employment has also forced the unions in the richer part of the world to reassess the concept of the collective bargaining process. For flexible workers, the unions cannot be arbitrators with regard to job security and higher wages. Nonetheless, unions can provide a representative for flexible workers to negotiate minimum guarantees that will serve as a floor for individual bargaining. As the AFL-CIO stresses, in the present era unions develop and put into effect multiple models, representing the needs of different groups of workers.[56]

In Canada, likewise, in order to break out of its traditional male image, the trade union movement has pledged to be a watchdog for those who cannot defend themselves.[57] To attain this goal, unions are placing greater emphasis on community-conscious social unionism, building close associations with anti-poverty lobbies, women's groups and the churches.

The new direction in unionism, as pursued in the affluent world, warrants some caution, if it is to be emulated in poorer countries. In citing experiments in the developed world, my aim is to highlight the creative approaches towards mobilisation needed in the current phase of privatisation, deregulation and putting out of work. Trade unions in poorer parts of the world have far fewer resources; hence, it is not always possible for them to experiment or imitate strategies that have proved effective elsewhere. The problems of mobilising workers are not similar either; and it is important to be aware of the differences in the challenges facing unions as well as the commonalities.

The current attitude of the mainstream unions

In spite of some questioning, the existing structure and ideology of the mainstream unions understandably are not conducive to organising casualised workers. The membership of the unions comes primarily from the core workers of the organised sector; thus, casualised workers pose a serious threat to their effectiveness and power. The statements issued by unions frequently reflect their entrenched fear.

In 1989, a proposed lifting of bans on homework in sectors where it had hitherto been prohibited provoked a strong response from the unions in the United States:

> Legalised industrial homework will mean a return to unchecked exploitation of desperate immigrant workers; a return to the industrial dark ages of our history when the bosses wrote their own rules, the government shrugged as if helpless, and ordinary working men and women suffered cruelty, a return to a meaner, more brutal America which all of us long ago hoped we would not see again. Homework is an inherently exploitative system of labour, because it is a cheap way for employers to expand production.[58]

Homeworking, quite naturally, appears as a threat to the power of the organised workers. At times, it is also viewed by unions and management as a transitory phenomenon which will disappear with economic growth, automation and stricter surveillance.[59] The evidence, however, does not justify such a view. Homeworking has not disappeared or been abolished; some national trade union bodies, accordingly, have changed their policy and recruitment measures. The Trades Union Congress (TUC) in the United Kingdom has explicitly acknowledged homeworkers as valid employees whose welfare should be the concern of trade unions. In *Homeworking: A TUC Statement* (March 1985), the British unions are urged to include terms and conditions for homeworkers in collective agreements. Trade unions are also being encouraged to examine their structures to see if they are conducive to active participation by homeworkers (Appendix III).

The role of women's organisations

In changing the consciousness of the mainstream unions, women's organisations have played a crucial role. The Vrouwenbond van de Federatie Nederlandse Vakbeweging (FNV) – a women's union affiliated to the Federation of Dutch Trade Unions – has admonished the male trade unionists:

> Homework is in practice 'women's work'. Usually the most poorly paid types of homework are carried out by women. . . . The plea for abolishing homework has not led to the necessary improvement of their position. . . . The FNV should put more effort into improving the position of homeworkers through legislation and collective agreements, and into limiting homework through an active trade union policy.[60]

In the United Kingdom, The National Homeworking Group, an umbrella organisation of homeworking groups, has likewise stressed the importance of an integrated approach in improving the homeworkers' working conditions. In 1984, in consultation with the representatives of homeworkers, local authority workers, community groups and trade unions, the group, in a national conference, adopted a Homeworkers' Charter. The charter urged trade unionists and policy makers to be aware of the complex strategies that are needed to empower the homeworkers (Appendix IV).

In India, likewise, SEWA has been active in lobbying the national and international bodies, so that they acknowledge explicitly the status of home-based workers as valid employees and give them a proper share of the national welfare measures. It is partly because of the relentless campaigning of these women's groups that the International Confederation of the Free Trade Unions (ICFTU) adopted in 1988 a resolution on home-based workers in their 14th World Congress in Australia. In it, ICFTU makes a plea to the affiliated organisations to develop special mobilising programmes that are directed mainly at home-based workers (Appendix V).

In addition the ICFTU has committed itself to undertake action-oriented research in this area. The recent report, *On Organising Workers in the Informal Sector* (1990), is an initial step in this direction.[61] The International Labour Organisation is currently working towards formulating an improved legislative framework for homeworkers in all countries. The European Commission, likewise, is exploring the position of homeworkers explicitly in all discussions on atypical work.

Women's organisations have also been active in establishing international links between homeworkers' groups. In view of the ease with which home-based jobs can be moved from one region to another, it is important to have a global dimension to mobilising programmes for home-based workers. The EEC-funded Industrial Restructuring Education Network of Europe (IRENE) of Holland and the National Homeworking Group of the United Kingdom have been engaged in establishing an international network of researchers and lobbyists in this field. The Self-Employed Women's Association or SEWA of India has played a significant role in establishing links in informing public opinion.

Alternative trading organisations

A new countervailing power has emerged in the west which also acts as a watchdog for the implementation of protective legislation for casualised workers. It consists of a new breed of marketing and retailing companies which have set up fair trade with the producers of the Third World. These companies are called alternative trading organisations or ATOs. The umbrella organisation of ATOs is the International Federation of Alternative Trading (IFAT); its membership comes from forty western organisations from North America, Europe, Australia and Japan. The products sold by the ATOs include craft products, textiles, foodstuffs, tea and coffee. A large proportion of them are produced by home-based poor women.[62]

The principles of buying are clear for these organisations. They buy from those groups in the Third World that utilise the labour of home-based and casualised workers ethically. The subcontractors of the ATOs are by no means the cheapest; but they adhere to certain social as well as financial criteria. ATOs, in turn, help these groups with marketing, stock control and design advice. In addition, ATOs spend money in raising consciousness among the consumers about the use of exploited labour for the commercial world.

Compared with the mainstream retail companies, the power of the ATOs is still small. But they do include organisations such as Oxfam, which is one of the largest retailers in the UK. A recent market study undertaken by One World Trading Company in the US revealed that this ethical market now reaches 10 million people in North America. Such people are motivated not only by what the dollar buys for them, but also by what their dollar does in the producing community.[63] The growth of 'ethical and green consumerism' has given a new boost to the morale and market power of the ATOs in the last five years; it has encouraged some of them to take up seriously the role of watchdogs.

Traidcraft, the second biggest ATO of the United Kingdom, is one such example. Besides marketing its 'people-friendly' goods, it actively investigates international and national sourcing of the mainstream retailers. The innovative advertisements of ATOs appeal to the conscience of committed consumers. They also put moral and financial pressure on the big retailing companies to place orders where the countries' protective laws are observed (Appendix VI).

The alternative trading organisations receive relevant information from research undertaken by the non-governmental organisations and pressure groups of the developed world. SOMO (Stichting Onderzoek Multinational Corporations) in the Netherlands and Women Working Worldwide in the UK are two such organisations well-known for their investigative work.

A case for planning from below

Committed research undertaken by these intellectual activists has meant that the role of informal work in the wealth creation of a nation now receives increased attention. This has contributed to a fundamental reassessment of accepted methods of collecting statistical information. It may be expensive and time-consuming to devise alternative ways of gathering data, but it is now generally recognised that there has been a serious undervaluation of women's labour in the Gross National Product.

The new awareness also highlights the need to change the trajectory of current development plans. So far, most of the national and international funding has gone to projects that have been devised and executed by professional experts. These are generally macro-projects, aimed to increase the rate of growth of a country's per capita income. The goal of such projects needs no justification. Indeed, without an overall improvement of the economy, it is difficult to improve the working and living conditions of casualised workers only through the empowerment of grassroots organisations. Yet the fact remains that the macro-projects generally benefit the male workers of the organised sector. The benefits of even the most successful projects, such as 'the green revolution', have systematically eluded women. On the grounds of distributive justice, it seems necessary, therefore, to channel resources to the kinds of organisations that are now being initiated and sustained by women workers of the unorganised sector.

This directional change in developmental funding is likely to prove cost-effective as well. The priorities of women workers' organisations are predominantly shaped by the deeply felt needs of the families and the community of the recruited members. Outside consultants rarely play an important role in deciding their strategies and programmes. By empowering these community-based organisations, the agencies are therefore likely to reap a high social and economic return on their funds. It is not only the casualised workers who work in isolation, it is also the women activists who engage, against all odds, in mobilising them. These activists will benefit enormously if they are given a chance to exchange experiences and learn about the strategies of their counterparts in other regions of the world.[64]

APPENDIX I: NINETEENTH OF SEPTEMBER UNION MEETING FOR INTERNATIONAL EXCHANGE AND SOLIDARITY

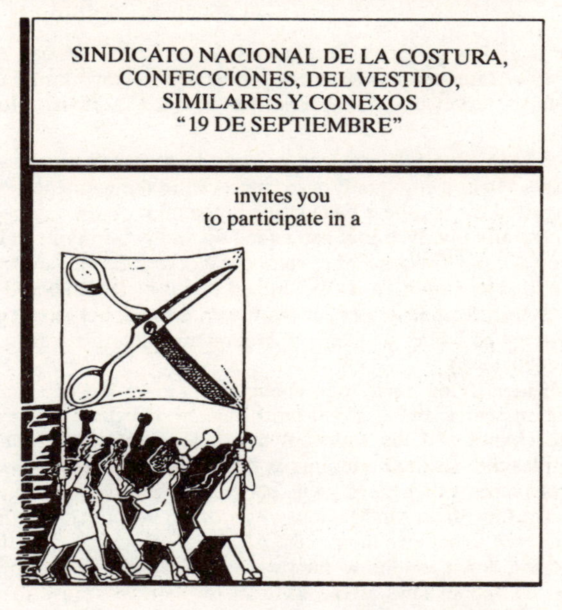

MEETING FOR INTERNATIONAL EXCHANGE AND SOLIDARITY

– to be held in Mexico City, October 22–25, 1986 to promote national and international exchange, analysis and solidarity.

Participants will include women from the September 19 National Garment Workers Union, and Mexican, North American and European women involved in women's labor and international education and organizing.

The program includes both internal meetings for participants to exchange experiences and develop strategies, and open forums to explore the issues with the wider public.

Meetings will be held at the union's headquarters in the San Antonio Abad garment district in Mexico City, and participants will stay as guests in the homes of union members. All meetings will be simultaneously translated into English and Spanish.

To register, write to: Cornadas de Intercamio
Apdo 12–709
Col. Narvarte
Mexico 03026
MEXICO

Include information about yourself, your organization and write about what expectations you bring to these meetings. As soon as possible, send us your arrival information (time, airline, flight #). A $20–50 donation to the union is requested to cover costs – please bring this with you in cash. We need your registration by October 8.

If you send us $2.20 in US stamps, we can send you more detailed information on the history of the union.

Allow 2–3 weeks for response to all mail.

APPENDIX II: ORGANISATION AMONG MEXICAN CLOTHING WORKERS IN EL PASO

Excerpts from an interview given by Cecilia Rodriguez, Centro Obrero, El Paso, Texas, USA, in June 1989 at International Restricting Education Network of Europe (IRENE) conference in Holland, reported in *News from IRENE*, No. 11, Tilburg, Holland, 1989).

I became involved in this work because of my own background as a second generation Mexican in El Paso. Most of my family were involved in the garment industry, and as a student I had supported the two-year Fada strike in the mid-1960s, and had seen how the women workers centrally involved in it had ended up totally burnt out and cynical, while the men took the credit. In 1981 some of us started a Workers Centre, and our first struggle was over the role of women within it. We argued that the centre should have a special focus on women as they formed 85 per cent of the workforce; but the men accused us of taking up the ideology of white 'women's liberation' and felt this was inappropriate for Mexican women (Chicanas).

The process of identifying goals took about three years. We eventually set up as a workers' information centre. Being a non-profit-making structure allowed us to receive funding from the church and the state but prohibited us from labour organising or lobbying. These rules did not limit our aims as the workers themselves had become very anti-union, both because of their negative experience of the Fada strike and because of their experience of yellow unions in Mexico, who worked in collusion with the employers and often used violence to achieve their ends. Also in 1983, many of the largest garment factories were closing down and the women were increasingly having to work in sweatshops. Lastly, the US labour laws are so difficult for workers to use that it is easier to operate outside them. For all these reasons, we were clear that we did not want to establish a union. The long-term aim we identified in 1981–2 was to build up committees of five or six women in each factory to address the problems of that particular factory.

We knew that if we were to win the trust of the workers and achieve these goals, we had to develop a strong leadership who were themselves textile workers. However, there was a lot of strife between staff over this issue: between middle-class Chicanas and factory workers. The middle-class women often felt that garment workers did not have skills, education or understanding to organise. This struggle was important in establishing the principle that the workers in the Centre could not see themselves as superior or different from the textile workers; the middle-class workers had to learn to live, eat and breathe conditions in the garment factories, to appreciate the workers' point of view. If you have a commitment to develop leadership, then you must understand that women have to go through stages of development; it is your responsibility to provide the information they need in order to make more strategic decisions. It was this understanding that finally determined the success of the group.

We undertook two activities in the factories that helped establish our image as a new group. In 1984 we undertook a 'factory gate' questionnaire survey about workers' needs: we asked about income, benefits, transport, etc., but nothing about employment conditions or unions. Then in 1985 we organised a health screening programme for 800 women in twelve factories. The contacts we made through these activities were followed up by home visits and a monthly newsletter. Our message was that we were 'women helping women' to develop alternatives, for ourselves and our daughters. It was not until 1985 that women finally began to bring their work problems to the Centre, and then we could in all honesty say, 'we have to organise, you cannot negotiate on your own'. We waited until the women themselves raised the questions.

We developed a leadership training programme over a number of years of trial and error. This includes the study of capitalism and political economy – concepts of wages, profit, use value, exchange value – concepts that workers need to understand in order to negotiate with employers. We also study labour history. Employers in El Paso try to make workers think that they are outsiders, privileged to be 'allowed' in. Studying labour history shows these workers that immigrant workers have always been at the forefront of the labour movement in the US. We provide as many opportunities as possible for women to learn about women's struggles in other parts of the world; through films, exchanges and visits. Paulo Freire's methods are used a lot: getting workers to draw and talk about their experiences in the factory.

However, much of the learning comes from the time spent in developing the organisation itself. We have five women on the Board of Directors of the Centre, and it is this group who are involved in the training programme, and in the policy making for six hours a week. They are learning all the time through the process of decision making: setting up factory committees, recruiting new members, and developing activities and services for members. They have full responsibility for the membership fund, part of which is used for a revolving emergency loan fund. They have developed a food-buying club for cheaper food, an advocacy service, English classes and immigration advice. They are also involved in the development of factory committees and work with women in the factories to document conditions.

This group of women share the information learnt in the training programme with the members who come into the Centre by setting up an exhibition or wall newspaper each week. The role of the women both domestically and in the workplace is a frequent theme in these; and they are around to explain it, so that every woman who comes into the Centre for food or services, also gets a political message. So all the local organisation is now done by local organisers; they are very strong women and it will be difficult to defeat them.

APPENDIX III: EXCERPTS FROM THE TRADE UNION CONGRESS (UNITED KINGDOM) HOUSE STATEMENT ON HOMEWORKING

Unions should seek to achieve employee status for homeworkers, with the company concerned making deductions for PAYE, national insurance contributions, and employer and employee pension contributions.

Trade unions should seek to include terms and conditions for homeworkers in collective agreements.

Trade unions should obtain from employers details of the volume and type of work put out to homeworkers . . .

Unions should remind safety representatives that, under the *Code of Practice on safety representation*, employers are required to make available to them information on articles or substances which an employer issues to homeworkers.

Unions may wish to consider setting up groups or committees in companies using homeworking to monitor the volume, type and frequency of working being put out and the terms and conditions of homeworkers.

The names and addresses of homeworkers should be obtained from the employer.

All union publications, leaflets, etc., whether national, local or in-company, should be sent to homeworkers.

Contact should be made with homeworkers wherever possible. Unions may wish to recommend to lay representatives from the company concerned that they carry out this task.

Contact with homeworkers can be made through links with local authority homeworking officers and local support groups where these support trade union membership.

Trade unions will need to examine their structures to see if they are conducive to homeworkers, particularly women, playing an active part in them. This examination would be in line with the TUC recommendations contained in the TUC publication *Equality for Women within Trade Unions*. Unions may also wish to consider establishing separate sections to provide a special identity and to cater for homeworkers' specific problems.

Unions may wish to make an official responsible for the recruitment of homeworkers and to conduct special recruitment campaigns.

All trade unions, whether organising in homeworking industries or not, should seek to make their members aware of the problems of homeworkers and of the potential for the spread of homeworking.

Source: *Homeworking: A TUC Statement*, London, 1985

APPENDIX IV: HOMEWORKERS' CHARTER IN THE UNITED KINGDOM

The demands contained in this Charter are those made by Homeworkers. The vast majority are women who suffer the triple burdens of childcare, housework and paid employment. Homeworkers are caught in the poverty trap and as such provide cheap, unorganised labour, especially for the sectors of industry which perpetuate the worst employment practices. Homeworking, especially in the new technology industries, both in manufacturing and the provision of services, is on the increase; it is now being promoted as the way of working in the future even by multi-national concerns. It is clear that the bad employment practices of traditional industries are being imported into the newer ones to the detriment of worker organisation. Homeworkers, who are particularly vulnerable to racist and sexist exploitation, subsidise their employer's profits and there is no doubt that given better opportunities few Homeworkers would work at home.

This Charter therefore demands that:

1. FREE ADEQUATE CARE OF DEPENDANTS IS AVAILABLE FOR HOMEWORKERS. A majority of homeworkers say that they are forced to work at home in order to look after children, or sick, elderly or disabled dependants, and that if adequate care were freely available this would enable them to work outside the home.
2. RESOURCES ARE PROVIDED TO ENABLE HOMEWORKERS TO MEET TOGETHER FOR MUTUAL SUPPORT, ORGANISATION AND CAMPAIGNING. Homeworkers live and work in isolated conditions with little or no opportunity for exchanging information with each other, or for recreation. If homeworkers are to improve their economic status these resources must be made available.
3. EMPLOYEE STATUS IS GIVEN TO HOMEWORKERS. Lack of clarity about the employment status of homeworkers has resulted not only in the casualisation of homeworkers' labour but also in the loss of other rights and benefits which depend on proof of employment status, e.g. Sick Pay, Unemployment Benefit, Maternity Benefit, Family Income Supplement, Pensions, etc. In addition, homeworkers subsidise their employer's business by paying rent, rates, heating, lighting, running and maintaining their machines. The employer also does not pay any staffing costs, thus avoiding capital and revenue outlay.
4. AN END TO RACIST AND SEXIST PRACTICES AND THE REPEAL OF RACIST AND SEXIST LEGISLATION. The isolation and fear homeworkers suffer are compounded by the laws, attitudes and practices of a society which is essentially racist and which denies the right of all women to participate informs the attitudes and procedures which exclude women and black and minority ethnic people from the benefits of the community to which they contribute.
5. THE ADOPTION OF A NATIONAL MINIMUM WAGE. The adoption of a national minimum wage for all homeworkers is essential in order to end the super-exploitation of homeworkers, people with disabilities and other unprotected groups. One national minimum wage would eliminate the problems associated with the complicated Wages Council Orders and their present lack of enforcement.
6. THE AMENDMENT OF RELEVANT REGULATION TO ENSURE THAT HOMEWORKERS AND THEIR FAMILIES DO NOT SUFFER INJURY, DISEASE OR SICKNESS AS A RESULT OF THEIR WORK. Homeworkers use dangerous substances such as glues, fixes and solvents, unguarded machinery and VDU's in their homes without the protection afforded all other

workers. They carry the responsibility of the health and safety of themselves and their families which should by right be that of their employer. The Health and Safety at Work Act must be amended to include all homeworkers.

7. COMPREHENSIVE TRAINING AND EDUCATION OPPORTUNITIES FOR HOMEWORKERS. Given the opportunity homeworkers prefer to work outside the home. Some lack the necessary skills and education to participate in the labour market; some are skilled in one process or production which may well become obsolete in a rapidly changing industry; some skilled workers may have been out of paid work while raising children and their skills need upgrading; some have never had the opportunity.

Source: *Report of the 1984 National Homeworking Conference*, The Greater London Council, London, 1985

APPENDIX V: THE INTERNATIONAL CONFEDERATION OF FREE TRADE UNIONS ON HOME-BASED WORK

The 14th World Congress of the International Confederation of Free Trade Unions, meeting in Melbourne from 14 to 18 March 1988;
NOTES
that home-based workers constitute an important part of the working population in many non-industrial and semi-industrial countries, where their conditions of work and remuneration amount to some among the worst forms of exploitation;
that policies promoted by conservative governments, such as the degradation of the labour market, have resulted in an increase in home-based work also in industrialised countries, sometimes in relation with new technology such as computerisation, and that these workers in most cases lack social protection;
that a great majority of both old and new home-based workers are women and that this fact is related to their weak position on the labour market and the lack of possibilities to combine regular work with family responsibilities;
CALLS FOR
a world-wide census of home-based workers, for the adoption of international labour standards and for national legislation in all countries where significant numbers of home-based workers are employed guaranteeing their basic rights in terms of working conditions, wages and welfare;
CALLS ON
its affiliated organisations to develop special organising programmes directed at home-based workers, whenever appropriate in conjunction with their women workers' programmes, to demonstrate in their actions and policies practical solidarity with trade union organisations of home-based workers where they exist, to press for legislation protecting home-based workers and to resist economic policy decisions that destroy the livelihood of home-based workers and to give them access to collective bargaining;
RESOLVES
to promote research on the legal, economic and social conditions of home-based workers and to publicize its results.
Source: *Decisions of the 14th World Congress*, ICFTU, Brussels, 1988

APPENDIX VI: INNOVATIVE ADVERTISEMENT BY A UK-BASED ALTERNATIVE TRADING ORGANISATION

How Clean Are Your Clothes?

Would you buy clothes that:

- were made by a UK homeworker paid just 50p an hour?
- came from a Philippine sweatshop where unions are banned?
- were made by a woman locked into a Bangladesh factory 23 hours a day?
- came from a factory where cotton dust causes lung disease?

The trouble is, none of us know how 'clean' the clothes are that we buy; there is a conspiracy of silence about where and how they are made. Turn over the page to find out how you can help to break it.

NOTES

1 Marilyn Waring, *If Women Counted: A New Feminist Economics*, Macmillan, London, 1989.
2 Phil Mattera, 'Small Is Not Beautiful: Decentralized Production and the Underground Economy in Italy', *Radical America*, 14(5), 1980.
3 The Economics Foundation, *The Charter for Economic Democracy and Sustainability*, Universal House, 89–94 Wentworth Street, London E1, 1989.
4 'Role of Employers' Organisations in the Informal Sector', Background paper for the ILO South Asian Employers' Symposium at Colombo, Sri Lanka, 21–24 March, ILO, Geneva, Bureau for Employers' Activities, 1989, pp. 6–7.
5 Swasti Mitter, *Common Fate, Common Bond: Women in the Global Economy*, Pluto Press, London, 1986.
6 Swasti Mitter, op. cit. pp. 112–15; Fiorenza Belussi, 'Benetton Italy: Beyond Fordism and Flexible Specialisation', and Pauline Conroy Jackson, 'Homeworking in Italy in the Age of Computer Technology', in Swasti Mitter (ed.) *Computer-aided Manufacturing and Women's Employment*, Springer-Verlag, London and Berlin, 1992.
7 Sheila Rowbotham, 'Causalization', *Zeta Magazine*, July/August 1989, p. 62; Andrew Sayer and Richard Walker, *New Social Economy: Reworking the Division of Labour*, Blackwell, Oxford, 1992, Chapter 5; Stewart Clegg and Fiona Wilson, 'Power, Technology and Flexibility in Organisation', in John Law (ed.) *A Sociology of Monsters: Essays on Power, Technology and Domination*, Routledge, London and New York, 1991.
8 Quoted in Rowbotham, 'Casualization'.
9 *New York Times*, August 1 1988, and quoted in Cynthia Enloe, *Bananas, Beaches and Bases: Making Feminist Sense of International Politics*, Pandora, London, Sydney, Wellington, 1989, p. 159.
10 SEWA, *My Home, My Workplace*, Report of the National Workshop on Home-based Piece-rate Workers, Ahmedabad, 17–19 November 1987, Ahmedabad, India, p. 47.
11 A. Prasad and K.V.E. Prasad, *Homeworking in India: A Review*, Monograph prepared for the ILO, Geneva, 1987 (mimeo), p. 57.
12 Anne Posthuma, 'Japanese Techniques in Brazilian Automobile Components Forms: Best Practice Model or Basis for Adoption?', Paper presented to the conference 'Organisation and Control of the Labour Process', Aston University, UK, 28–30 March 1990, pp. 23–7.
13 Rosalinda Pinenda-Ofreneo, op. cit.
14 Nirmala Banerjee, 'The Unorganised Sector and the Planner', in A.K. Bagchi (ed.) *Indian Economy, Polity and Society*, Oxford University Press, 1988, p. 91.
15 Nirmala Banerjee, ibid.
16 I.S.A. Baud, *Women's Labour in the Indian Textiles Industry*, Tilburg University, the Netherlands, 1983, Chapter 2, p. 23.
17 I.S.A. Baud, *Forms of Production and Women's Labour: Gender Aspects of Industrialisation in India and Mexico*, Technical University of Eindhoven, the Netherlands, 1989.
18 Bela Banerjee, 'Working Women' (Kajer Meyera), *Parichaya*, December 1977, Calcutta, p. 64.
19 UNIDO, *Industrial Development in Developing Countries: Trends and Perspectives*, 1989.
20 Interview by Swasti Mitter, 1989.
21 Maria de los Angeles Crummet, 'Rural Women and Industrial Homework in Latin America: Research Review and Agenda', WEP Working Paper, WEP10/WP 46, ILO, Geneva, June 1988, p. 11.
22 Mario Vargas Llosa, 'The Silent Revolution', *International Health and Development*, 1 (1), March/April 1989, p. 7.

23 ibid., p. 15.
24 Marja-Liisa Swantz, 'The Effect of Economic Change on Gender Roles: the Case of Tanzania', *Development*, Journal of the Society for International Development, 1988, Volume 2/3, p. 94.
25 Interview by Swasti Mitter.
26 Banerjee, 'Working Women'.
27 Crummett, 'Rural Women and Industrial Homework in Latin America', p. 13.
28 S.E.G. Perera, *National Monograph on Homework in Sri Lanka*, prepared for the Working Conditions and Environment Department, ILO, Geneva, 1987 (mimeo), p. 27.
29 Cynthia B. Costello, 'Home-Based Employment: Implications for Working Women', The Women's Research and Education Institute, 1700, 18th Street, NY, USA, 1987, p. 7.
30 Roeland Van Geuns, 'An Aspect of Informalisation of Women's Work in a High-Tech Age: Turkish Sweatshops in the Netherlands', in Mitter, *Computer-aided Manufacturing and Women's Employment*.
31 Swasti Mitter, 'Industrial Restructuring and Manufacturing Homework: Immigrant Women in the UK Clothing Industry', *Capital and Class*, 27 November, Winter 1986.
32 ibid.
33 Adrianna Marshall, 'Non-standard Employment Practice in Latin America', International Institute for Labour Studies, Discussion Paper, Geneva, 1987.
34 ibid., p. 31.
35 ibid., p. 32.
36 Kattiya Karnasuta, *Homeworking in Developing Countries: A Case of Thailand*, prepared for ILO, Geneva, 1987 (mimeo), p. 19.
37 Rukmini Rao and Sahba Husain, 'Invisible Hands: Women in Home-based Production in the Garment Export Industry in Delhi', in Andres Menefee Singh and Anita Kelles-Vitanen (eds) *Invisible Hands: Women in Home-Based Production*, Sage Publication, New Delhi and London, 1987.
38 *Breaking Down the Wall of Silence*, International Confederation of Free Trade Unions, Brussels, 1986, pp. 3–14.
39 Rosalinda Pineda-Ofreneo, 'Industrial Homework in the Philippines', prepared for the International Labour Office, Geneva, 1987, p. 30 ff.
40 *Pixote*, dir. Hector Babenco, Brazil, 1981; *Salaam Bombay*, dir. Mira Nair, 1988.
41 Hernando de Soto, *The Other Path*, Harper & Row, United States, 1989; also, 'An interview with Hernando de Soto', *International Health and Development*, Vol. No. 1, 1989, p. 10 ff.; Llosa, 'The Silent Revolution'; p. 19.
42 Interview by Swasti Mitter on 7 April 1989 at the Regional Organisation Committee of CISL, Lucca, Italy.
43 Cynthia B. Costello, *Home-based Employment: Implications for Working Women*, Women's Research and Education Institute, NY, USA, 1987, p. 7.
44 Prasad and Prasad, *Homeworking in India*, p. 102.
45 ibid., p. 103.
46 Quoted in *Shramshakti: Report of the National Commission on Self-Employed Women and Women in the Informal Sector*, New Delhi, 1989, p. 255.
47 *Shramshakti: Report of the National Commission on Self-Employed Women and Women in the Informal Sector*, New Delhi, 1989, p. 225.
48 ibid., p. 128ff.
49 Prasad and Prasad, *Homeworking in India*, p. 109.
50 Cynthia Enloe, op. cit., p. 173.
51 Interview with Swasti Mitter, June 1990.
52 *The Changing Situation of Workers and their Unions*, Report by the AFL-CIO Committee on the Evolution of Work, February 1985, p. 18.
53 'TGWU Unveils New Face of Unionism', *Guardian*, 27 February 1987.

54 'Unions Want to Champion the Needs of Women Workers', *Guardian*, 26 May 1986.
55 *International Labour Reports*, 31, January/February 1989, pp. 7–9.
56 *The Changing Situation of Workers and their Unions*, op. cit., p. 19.
57 *International Labour Reports*, op. cit. p. 8.
58 *Conditions of Work Digest*, 8(2), ILO, Geneva, 1990, p. 221.
59 See, for example, Resolution No. 2, concerning industrial homework in the clothing industry, adopted by the tripartite technical meeting for the clothing industry, Geneva, 1964.
60 *Conditions of Work Digest*, p. 192.
61 Swasti Mitter, *On Organising Workers in the Informal Sector*, ICFTU, Brussels, 1990.
62 *The Network*, newsletter for the equal exchange of information on trade and technology, July–September 1989.
63 *The New Internationalist*, February 1990, pp. 24–5.
64 Sheila Rowbotham, *Homeworkers Worldwide*, Merlin Press, London, 1993.

Chapter 2

Women in the Bombay cotton textile industry, 1919–1940

Radha Kumar

ABSTRACT

Women played a crucial role in the phase of industrialisation in India which began in the nineteenth century. They were drawn into certain occupations like the mechanised textile industry in Bombay. Radha Kumar's account shows the existing sexual division of labour was one factor in the composition of the new labour force.

By the 1920s women were at once industrialised and often militant factory workers, yet linked to casual employment networks in their communities. Marginalised within workers' unions and defined by reforming policies that they were unable to affect, their aspirations and interests in the 1920s and 1930s can be deduced from their resistance to rationalisation which increasingly was forcing them out of the mills and waged employment.

This historical perspective enables us to see that the sexual composition of a labour force is neither constant nor predetermined and that even within forms of factory production various degrees of casualisation have persisted, especially in connection to women workers.

INTRODUCTION

In the early phases of industrialisation in India, women constituted an important source of labour for textile factories, jute mills and mines. Almost a quarter of the workforce of the Bombay cotton textile factories, for example, were women, from the mid to late nineteenth century, when the factories multiplied, until the 1920s, when rationalisation and retrenchment set in. Women had traditionally been employed in textiles in domestic or handicraft forms of production. This work had consisted largely of preparatory processes: cleaning and sorting cotton, processing it for spinning, or spinning it itself. Women weavers were a rarity. The factory system adopted this traditional division of labour and added to it. Because the first factories in Bombay were spinning factories, spinning itself became the kind of skilled occupation which was reserved for men. Though by the early twentieth century the Bombay cotton textile industry expanded into weaving as well as spinning, this division of labour persisted, and there were neither women weavers nor women spinners in the Bombay mills.

The first mills were built in Bombay in the 1860s. Between 1860 and 1908 eighty-five cotton textile mills were built, mostly in the period 1860–90. From 1890 to 1922 fewer new mills were built, but the old ones enlarged their production from spinning to weaving, and also shifted to finer counts of yarn. The years of World War One saw a boom in the industry, and profits were especially high between 1917 and 1922. In the early 1920s increasing competition in the international market, especially from Japan, and from up-country mills in the domestic market, caused a slump in the industry which was worsened by the world-wide depression which followed. A series of measures were undertaken to reorganise the industry from the late 1920s through the 1930s, ranging from closures to changes in work organisation, lay-offs and wage cuts. The situation, however, only really improved during World War Two, when demand was increased by military requirements and the curtailment of foreign competition.[1]

The mills occupied over a quarter of Bombay island, spreading from the centre of the city into the far north. One-ninth of the city's population were cotton mill workers, most of whom lived huddled together around the mills in ramshackle, privately owned tenements, or *chawls*. The separation of home and workplace, thus, was marginal, and workplace politics often spilled over into neighbourhood politics.

Most of the workers were migrants from neighbouring districts in Maharashtra, who had been recruited by labour contractors (commonly known as jobbers or *mukaddams*), through kin, caste and regional networks. Until the 1930s, the jobber controlled the labour force and the organisation of work; after this period, fitful attempts were made to curtail his powers.[2] Though a large proportion of the workforce were single male migrants, entire families were also recruited into the industry, often through extended family connections (aunts, uncles, cousins, in-laws).

Over a third of all workers in the Bombay cotton textile industry came from one particular district, Ratnagiri, a poor and over-populated coastal area. And roughly two-thirds of the women employed in the industry came from this district,[3] where sexual segregation appears not to have been as rigidly practised as elsewhere, partly because the sex ratio favoured women and partly due to poverty.[4]

Most of the women workers in the Bombay cotton textile industry were winders and reelers. Over 90 per cent of winders were women, and reeling departments were entirely staffed by women. The work of a winder entailed removing the yarn from bobbins which came off the ring frames, and rewinding it into convenient lengths for the loom. Threads which would go into the weft of the cloth were wound on to pirns, small bobbins which fitted into the shuttle; threads which went into the warp were wound on to larger cylindrical bobbins called cheese or cone. Reeling was part of the process of producing yarn rather than cloth: reelers wound yarn into hanks, which were then tied into bundles and packed into bales for sale. Until the early twentieth century most winding was done by hand, while reeling was partly mechanised. The majority of women in both departments were pieceworkers.

The mills were characterised by a seasonal and fitful pattern of employment, expanding and contracting their labour force as well as working time according to short-term demand and production imperatives. A vast substitute or exchange labour force was kept on hand (known as *badli* labour), and there were thus relatively few full-time workers. Partly due to this, wages were at the margin of subsistence, though records indicate that they lagged behind prices in any case.[5] Women were particularly affected by this pattern of employment: a 1926 Labour Office survey into hours of work showed that the lowest attendance figures in the industry were for the winding and reeling departments. Only 20 per cent of reelers and 23 per cent of winders worked full-time.[6]

In 1895 women constituted almost a quarter of the labour force.[7] As the mills expanded and the labour force increased, this proportion remained roughly the same. When World War One broke out there was a slight fall in numbers employed: demand fell short of production, and the mills worked short-time. During and after the war the industry boomed, and from 1918 to 1926 there was a steady increase in both the total labour force and the numbers of women employed. While the proportion of women to the total labour force in 1919 was the same as it had been in 1911, there was a 25 per cent increase in the numbers of women employed and by 1926 this went up to 60 per cent.

From the late 1920s, there was an overall decline in the employment of women in the Bombay mills, which occurred in two phases. Between 1926 and 1930, over 4,000 women lost their jobs, and between 1930 and 1939 another 7,000 women had been retrenched from the industry. These of course were the years of depression for the Bombay cotton mills, and when the mills began to recover, the numbers of women workers rose. However, the numbers of men rose far more sharply than those of women (see Figure 2.1).

Proportionate breakdowns tell a slightly different story. From 1911 to 1924 the proportion of women workers to the total labour force remained roughly a fifth, and from 1925 to 1931 their proportions tended to oscillate between a quarter and a fifth of the total. While the number of women employed daily was highest in 1926, their proportion to the total labour force was highest in 1930. After this there was a steady decline, and in 1947 women workers were only a little more than a tenth of all workers (see Figure 2.2).

The retrenchment of women from the Bombay cotton textile industry was linked to what was optimistically called its 'rationalisation measures', which were directed towards rescuing the industry from its slump during the depression. M.D. Morris has convincingly established that the mills did not adopt most of the proposed schemes for rationalising the industry: wages were not standardised. Though there was an increase in mechanisation, the quality and numbers of machines put into operation continued to vary enormously from mill to mill; while work was frequently intensified, it was not necessarily better organised;

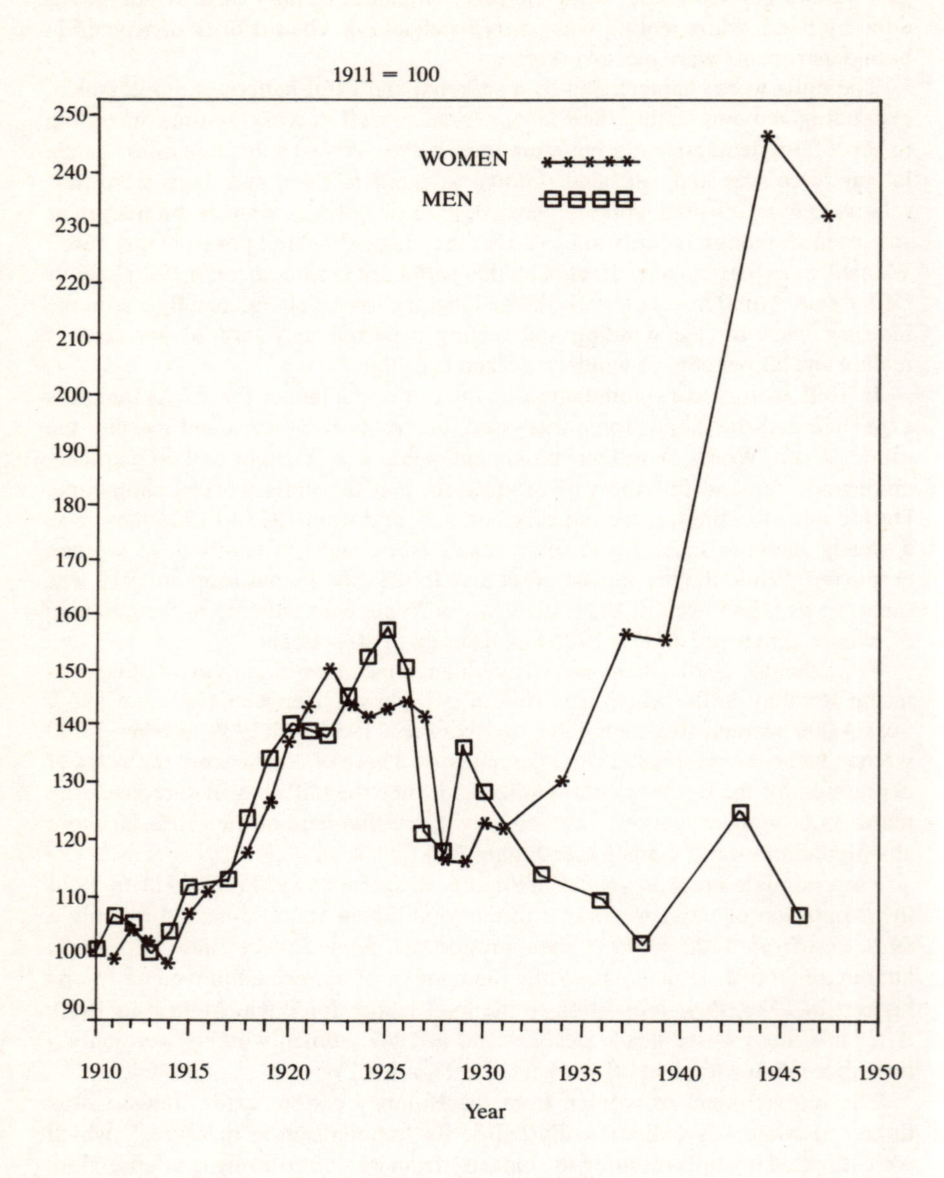

Figure 2.1 Average daily employment of men and women in Bombay cotton mills, 1911–47

Source: M. D. Morris, *The Emergence of an Industrial Labour Force in India*, Oxford University Press, Bombay, 1965, pp. 217–19

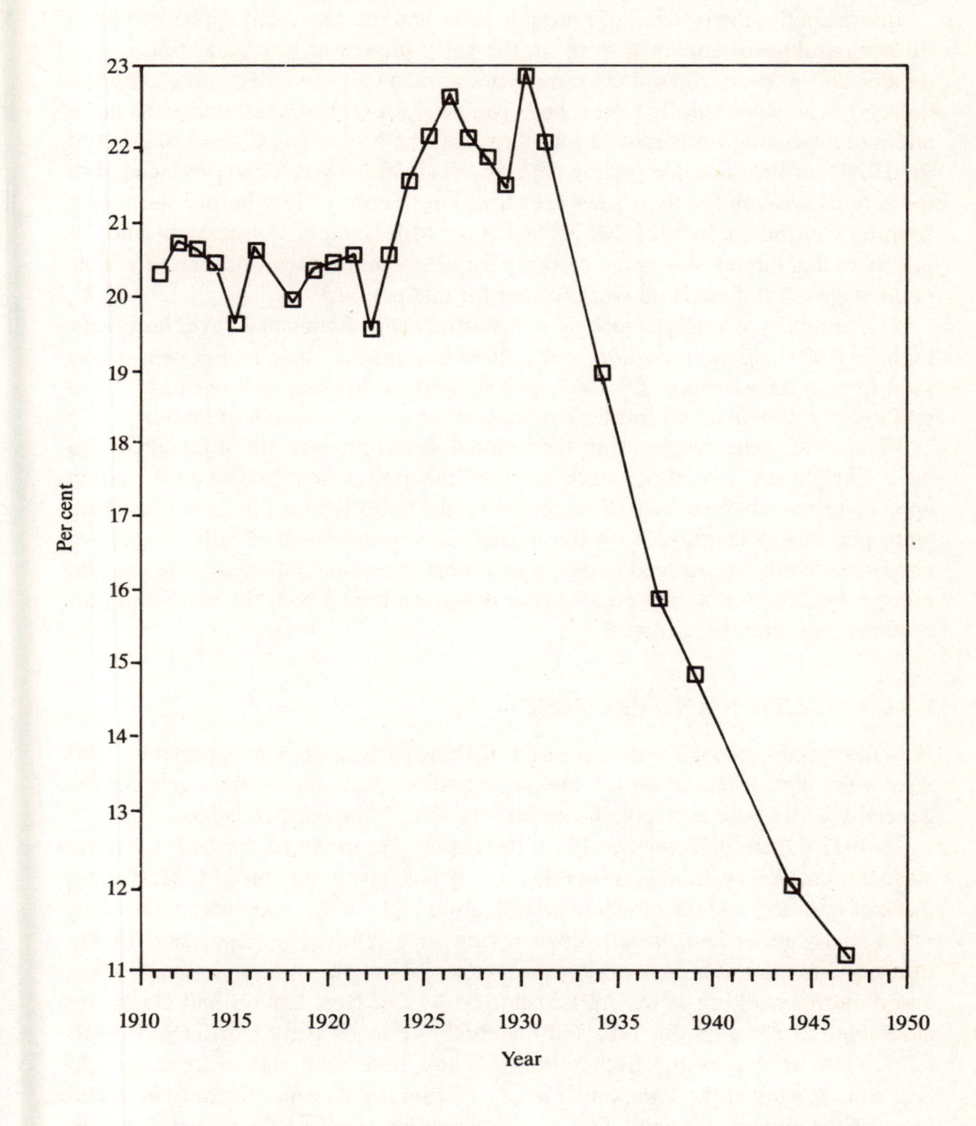

Figure 2.2 Proportion of women of total employed daily by Bombay cotton mills, 1911–47

Source: M. D. Morris, *The Emergence of an Industrial Labour Force in India*, Oxford University Press, Bombay, 1965, pp. 217–19

and a large *badli* labour force continued to be maintained.[8] Perhaps the most singular fall-out of rationalisation was the retrenchment of women.

Interestingly, the process of rationalisation and retrenchment appears to have further casualised women's work. In the early phases of rationalisation it was reelers, and women who did the same work as men (such as siders, *tarwallas* and doffers), who were chiefly retrenched. The need for reelers was reduced as more and more spinning mills moved into weaving; the loss of the Chinese market in the 1920s further curtailed reeling departments in mills which had produced yarn for sale as well as for their own use. Speaking about reelers before the Strike Enquiry Committee in 1928, Mr Stones of the Mill Owners' Association said: 'it is very casual labour. A woman working for nine hours at this wage can earn very good wages, but there is no work for her for that period.'[9]

The numbers of women working in departments with men had never been very high. In 1921, siders, *tarwallas* and doffers comprised some 14 per cent of the total female labour force. By 1937, this figure had dropped to 8 per cent.[10] The process of rationalisation further entrenched the sexual division of labour.

Finally, at some stage during the rationalisation process, the mills appear to have introduced a rotating work system (the *pali* system). The *pali* system appears to have been a way of rationalising the *badli* labour force: *pali* workers were permanent employees on the muster rolls, but instead of substituting for absentees, work was rotated between *pali* workers, who continued to be paid on piece rates. The mills, however, appear to have retained both the *badli* and *pali* systems, juggling them around.[11]

ORGANISATION AND PROTEST

How did women respond to the casual and shifting patterns of their employment, and what were their forms of protest and organisation? Material on these questions is generally both sparse and sporadic, but several interesting points emerge.

In 1913 Jehangir Bomanjee Petit, the deputy chairman of the Mill Owners' Association, observed, 'labour in this country has now commenced to realize the force of numbers and the power of combination'.[12] Late the next year, in October, when many mills had closed down owing to a temporary depression in the industry and the rest were working short-time, 'about one thousand men, women and children working at the mills managed by Greaves, Cotton and Company assembled at Parel on the 14th and marched to the company's office at Fort to demand their wages for September'.[13] They had been on strike since 30 September, against the company's policy of running its mills for only eighteen days of the month. It seems that a fairly lengthy process of negotiations then ensued: realising that the company was not going to agree to run the mills full-time, the workers had offered to return to work if they could work twenty-one days in the month, because 'they could not live in Bombay on the wages they would get if they worked only eighteen days'. The agents, however, refused.

The 1914 strike implies the practice of equal representation by sex. Unfortunately no details are given, so we do not know whether the representatives were chosen by departments or mill, or whether they were simply jobbers and *naikins* (women supervisors) involved in the strike. However, departmental or trade-based strikes appear to have always been a common feature: by 1920, newspapers were increasingly reporting weavers' and spinners' strikes and strike details from 1923 on show that a large number concerned one or another occupation alone, among which the three most militant groups seem to have been weavers, spinners and winders. According to S.A. Dange, who rose to prominence in the 1928–9 general strike, and has since been generally regarded as one of the fathers of the Indian trade union movement, women winders were more militant than women reelers because they had 'more bargaining power'.[14]

By the early 1920s, the number of strikes in which women were active was beginning to attract attention. In 1923 the *Labour Gazette* commented:

> It is of interest to examine the frequency and causes of strikes which have taken place among women operatives in the Presidency during the last two and a half years. Since the month of April 1921, there have been no less than 19 strikes in the Presidency, in which women workers were involved. . . . Of the 19 strikes, 12 were due to questions of pay and bonus, five to personal causes and two to other causes. . . . Of the total strikes, 16 occurred in mills in Bombay. . . . The results of the strikes show that 16 were settled in favour of the employers, two in favour of the employees and one was compromised.[15]

Women appear to have been active in about a fourth of all strikes and over a tenth of the total strikes in this period were 'women-only' strikes. Close to half the strikes in which women were active were in the winding department. One of the earliest strikes against mechanisation was by women winders at the Rachel Sassoon Mill, in March 1926. The mill management had installed new cheese-winding machines, which they asked the women winders to use. One hundred and forty-five women winders struck in protest. The strike lasted for five days, and the police were called in to remove the women, who were throwing bobbins around.[16]

The great general strike of 1928 against rationalisation was prefigured by a strike of women winders at the Jacob Sassoon Mill, which brought out the whole Sassoon group of mills. On the morning of 2 January 1928, women winders at the Jacob Sassoon Mill went on strike against a proposed reduction of between 1 and 4 pice per 10 pounds of yarn wound. By the afternoon they had been joined by the spinners, and the next day all workers in the mill were on strike. Spinners at the Spring Mill were already upset at the management's notice offering a 50 per cent increase to all siders who would work a whole frame instead of just one side, and when they heard of the strike at the Jacob Sassoon Mill, they downed tools and joined the strikers. (Most of the mills in Bombay were built close to each other, and it was common practice in the textile workers' movement for strikers to go from mill to mill bringing workers out.) On the 2nd itself, strikers from both mills went to the Apollo and Manchester Mills, and brought the workers there out on strike. By 5 January, five more mills in the Sassoon group had joined the strike, and by 23 January all mills in the group were on strike.[17]

Though both the moderate Bharat Textile Labour Union and the communist Girni Kamgar Union were involved in the January strikes, the general strike which began in April 1928 appears to have been the first time that unions attempted to organise women, and the Girni Kamgar Union appears to have been the first union to have decided to do so. Reports imply the decision was made in the context of how best to sustain the strike. At a meeting in late May 1928 at Nagu Sayaji's Wadi, the communist stronghold at Prabhadevi which became the headquarters of the strikers, while discussing the fact that many mill workers, both men and women, had been forced to return to work, Dange suggested using all-women pickets to stand at mill gates, and forming a team of women and children to go from house to house to collect donations for strike relief. In early June, a fillip to the idea of all-women pickets was given by the discovery that some of the union's male pickets had taken to chatting and drinking tea with mill staff and police constables, 'and what is most humiliating', said one leader, 'I saw some volunteers flirting with some women millworkers!'[18] Though teams of women and children did try collecting strike relief, donations were sparse;[19] one reason might have been that most charitable institutions were controlled by mill owners. In mid-June the Municipal Corporation debated a motion to provide relief for the wives and children of strikers moved by a municipal councillor, Choksey; seconding it, Dwarkadas (a leading reformer, active with the textile workers) said, 'he was not asking for mercy on behalf of the working women concerned who had sufficient self-respect not to ask for alms'.[20] Several of the older women workers, however, had been forced to beg on the streets;[21] others were seeking work wherever they could find it, either in *bidi* (cigarette) making or in slipper making.[22] The latter was considered especially degrading as it entailed handling leather, regarded as work fit only for a member of the 'untouchable' caste because leather is categorised as polluting.

By the end of June, all-women pickets were stationed at several mills, armed with broomsticks to 'intimidate blacklegs'.[23] Reports have it that the use of women pickets was a counter-strategy to the mill owners' announcement that they would open the mills on 1 July; it seems the men were greatly amused by the idea.[24] A week later, at a jobbers' meeting a social boycott of mechanics, oilmen, clerks and others who were still working was urged, and the trade union leader S.S. Mirajkar suggested that pressure could be put on them through their wives and children.[25] It seems that though N.M. Joshi had got the police to allow picketing on condition that clerks and maintenance staff would be allowed into the mills unmolested, the Girni Kamgar Union had different ideas, and they were attacked both at mill gates and in their *chawls*.[26] Despite these efforts, it appears that it was the women's pickets which shamed these categories into staying away:

> Several classes of workers, like clerks and technicians who used to attend their respective mills and help to maintain the essential services stayed away from the mills for fear of being exposed to the indignity of being roughly handled, or at any rate insulted by militant women picketers armed with broomsticks.[27]

Though the suggestion of all women pickets had been made at the end of May, the pickets were eventually formed a month later, at the end of June. Contemporary accounts explained this delay as due to conventional caste-based disapproval of women's public activity. According to the *Times of India*:

> Labour leaders succeeded in enlisting only twenty-five female pickets on Monday morning, varying in age from 16 to 60. Only seven of their number actually came out for the purpose of picketing, and a very comical sight these naturally shy and retiring Maratha women presented in their new role, their sarees tucked up and wearing a red shoulder band. Their presence at the gates of two or three mills caused considerable amusement rather than excitement and drew large crowds.[28]

A few days later the same newspaper reported that labour leaders were finding it difficult to get any Maratha women strikers 'to consent to do the work of picketing', and had been forced to enlist the help of 'young women of the Mangarudi tribe',[29] at the price of 8 annas a day. It seems the women regarded the trade unionists as 'an over generous set of employers' while they 'for their part, seem to derive no little amusement in captaining their team of youthful pickets'.[30]

Despite the uniform tone of friendly and patronising derision which ran through these reports, they make clear the immediate and electrifying effect of the all-women pickets, which shamed many male non-strikers so deeply that they refused to go to work any longer. Moreover, within a few days of the all-women pickets being posted at the mill gates, one woman picket was assaulted by a policeman, while another was assaulted by a clerk.[31] There does not seem to have been any great hesitation in arresting women either: on 21 June a woman picket was arrested because some clerks complained that as they were going home along Prabhadevi 'she abused them and created a scene'.[32] A month later, on 31 July, a woman volunteer called Anasuyabai was arrested while picketing the Bombay Cotton Mill on the complaint of another lot of clerks. Clearly, the dignity of clerks (called *babus* in India) was most easily affronted. Anasuyabai was fined Rs 25 by the magistrate, a considerable sum in those days.[33] Reporting on women's pickets during the 1928 Jamshedpur steelworkers' strike, the *Times of India* said ten women had been arrested and rumour had it that women coolies were going to be dismissed for their activism.[34] The year 1928 appears to be when trade unionists awoke to the potential of feminine militancy.

How far were the *Times of India* reports accurate in saying that Maratha women were reluctant to picket, and so the lower-caste – *dalit* – Mang Garudis had to be brought in? Women mill workers, as we have seen, downed tools frequently, took part in all demonstrations, and could be violent. 'Shy and retiring' seems a somewhat inapposite description. On the other hand, picketing was a relatively new agitational tactic, and it exposed women to public gaze in a way in which demonstrations did not. Allied to some degree of customary disapproval of women's public activism, this might well have been a source of initial reluctance to picket among women mill workers. If they were wary of picketing in the 1928 strike, however, their hesitancy did not last long, for there were all-women pickets during the 1929, 1934, 1938 and 1940 general strikes, not to mention the individual mill strikes. Beginning work in the mills in the late 1920s, both Janabai Prabhu and Bikibai Guldekar, who were Marathas, said with pride that they had picketed at mill gates, and denied that there was any embarrassment attached to the act.[35] G. V. Chitnis of the Girni Kamgar Union said women pickets 'wore their red shoulder bands with pride', and often wielded their brooms against blacklegs.[36] 'Seeing a woman picketing,' said S.A. Dange, 'the strike-breakers would slink away.'[37] And according to S. V. Ghate, 'woman volunteers would stand in front of the mill gates. They said, "who amongst you is manly enough to get inside?" . . . Nobody would go inside.'[38] At the end of July, an additional means of shaming blacklegs was suggested: that pickets should hand sharp knives to men attempting to enter a mill, 'asking them to cut off their own noses instead of proving traitors to the camp'.[39]

The use of shame as a tactic is of course time-hallowed. We know that it was deployed by men as well as women within the textile workers' movement: earlier efforts at dissuading blacklegs had included blackening the faces of selected strike-breakers, generally those who were regarded as most influential.[40] And at some mill gates, strikers put up notices saying that blacklegs were 'the offspring of a harem'.[41] But women strikers were also used to shame male strikers. According to Chitnis, as the 1928 strike wore on, instead of continuing to 'defend the strike against strike-breakers' the men took to playing cards and drinking. Ushabai Dange (Dange's wife and a prominent activist of the union) held a big meeting of women strikers to discuss the problem: they resolved to march to the *chawls* and burn the playing cards, and the following day held a women's demonstration.[42] Describing a women's meeting held at the Damodar Thakersay Hall on 31 August, S. S. Mirajkar of the Girni Kamgar Union said at a men's meeting in Nagu Sayaji's Wadi that the women had decided to 'fight shoulder to shoulder' with the men, adding, 'should the men run away when the women say that they will beat the owners with blowing pipes?'[43]

Despite such initiatives, attitudes towards women's militancy were ambiguous. How complex feelings could be is best reflected in a speech made by A. A. Alwe, a mill-worker trade unionist, just under three weeks after Mirajkar's address. On 30 September, the Morarji Gokuldas Mill reopened on standardised wage rates, and became the first mill to break the strike.[44] Its manager, Ramsing Dongersing, had rough shelters built within the mill compound for the workers who rejoined, so that they could live on the premises and thus be protected from the strikers' ire. On the day that the mill reopened, there was a huge meeting of strikers at the gates, with a large number of women. Alwe's speech opened with compliments to the women, moved to threatening Dongersing in the name of both men and women, and then sharply veered to an expression of regret that women had come to picket:

> In the beginning I heartily congratulate the females who have attended this meeting. . . . The fact is that as soon as the rumour was afloat that the mill was started the male as well as the female strikers assembled in this meeting with a grim determination to fight and gain victory. . . . [Dongersing] must bear in mind that this is the last day of his life, and that men and women have assembled here to perform his funeral ceremony. . . . Another fact worth mentioning is that the females have assembled here for their question of bread. So they may go back to their homes. . . . It is not a happy thing that women should take part in any struggle so long as men are alive in Maharashtra and in India.[45]

Six months later the Morarji Gokuldas Mill was on strike again, and the strike was started by women colour winders. They had been complaining for some time about being given bad yarn which made their production fall from 20 to 15 pounds. They also complained that the jobber had asked them for bribes. The mill manager said the women were deliberately going slow, and ordered the gateman not to allow them into the mill. In a rare gesture, the other workers of the mill came out in sympathy with the women, possibly as a result of union pressure.[46] Two women pickets were arrested 'for disorderly behaviour';[47] a few days later they were acquitted 'for want of sufficient evidence'.[48] The Wadia group of mills was also on strike at this time, because the management dismissed 6,500 workers for going on a 'lightning strike' when the union had agreed that there would be no strikes without notice.[49] The refusal of the management to reinstate these workers was one major reason for the second general strike to occur in twelve months, which began on 26 April. But another, equally important, reason was that the mill owners picked out union volunteers for dismissal, and refused to record the evidence of the union on victimisation.[50] The mills had long used the tactic of dismissing strikers and employing new hands: in September 1923, one hundred weavers and winders on strike at the Bradbury Mill were dismissed, while in October striking weavers and winders at the Mathuradas Mills were discharged and other workers engaged; and in 1925 twenty women winders on strike at the Planet Mill lost their jobs.[51] By 1929, however, it seems that some mills even penalised relatives of militant workers: giving evidence before the strike enquiry committee which was set up in July, a woman called Parvati, employed in the Wadia-owned Textile Mills, said she had worked there for eleven years, but after the strike in March had been refused re-employment on the grounds that her son Devji was a union volunteer.[52]

In fact, strike and post-strike details indicate that throughout the 1930s the mills used strikes to retrench workers. The Rachel Sassoon Mill, for example, reduced wages of colour winders in mid-July 1933; the winders concerned, ninety women, went on a three-week strike; the management dismissed forty-one of them on the grounds that they wanted to retrench winders.[53] In March 1934 reelers in the Connault Mill struck work against the retrenchment of thirty-eight reelers,[54] and in October 1935 women winders went on strike against the retrenchment of a fitter.[55] In August 1939, 300 women winders at the Simplex Mills struck work against the proposed retrenchment of 106 winders, but when told by the Girni Kangur Union (Red Flag) that their strike was illegal under the newly passed Bombay Industrial Disputes Act of 1938, they returned to work.[56]

Women did not always see retrenchment as related to changes in industrial organisation. In 1936, the Labour Officer, A. W. Pryde, reported that a group of Kamathi women winders, who worked on the Universal winding machines, came to him complaining that they 'had been informed that all women of their caste would be discharged, and that a notice to that effect had been posted in the mill'.[57] They had been misled, but Mr Pryde does not say by whom. Dwarkadas adds the following:

> While I was working at the India United Mills a group of Konkani Kunbi women winders came to me and said they had come to know that all women of their caste were going to be dismissed. I told them they had been told a lie, but it shows how deep their fears go.[58]

Perhaps the most famous of women workers' strikes against rationalisation was the Phoenix Mill strike. On 31 March 1939, the management of the Phoenix Mill issued a notice that they would be closing down the mill from mid-April. Ten days later, they asked the workers of the ring frame and reeling departments to work an intensified load,[59] and when the workers refused, they announced they would close the mill that very day. The trade unionists G. V. Chitnis and Parvatibai Bhor say the strike was started by women reelers; according to Chitnis, the management had bribed one woman to agree to mind a whole frame by herself, instead of with another woman, as was usual: the other women assaulted her, and then went on a 'sit-down strike', blocking the stairs.[60] Eight hundred women workers occupied the mill compound from that evening (the mill was one of the largest employers of women); when the manager and the spinning master attempted to leave in a car, the women sat in front of the car, and the manager and spinning master were forced to spend that night in the office.[61] The next day the strike was called off, because both sides agreed to refer the dispute to the Industrial Disputes Court, under Justice Divatia. On the day the hearings began, the sixty-four people who were to give evidence were escorted to the Court by a huge procession, 'prominent amongst whom were a few thousand women'.[62] In May, Justice Divatia ruled in favour of the mill management;[63] two days later the Bombay government (now run by the Congress Party) issued a statement that the 'stay-in strike' was illegal. Legally, they said, remaining on the mill premises constituted unlawful assembly, criminal trespass and threat to property, and the government wanted to warn the workers that the mill owners were entitled to prosecute them for occupying the premises even after a settlement was agreed.[64] The statement was widely regarded as a 'betrayal' of the workers by a nationalist government.

Despite Justice Divatia's ruling the Phoenix workers remained on strike. When, on 19 May, the management attempted to bring two lorryloads of blacklegs into the mill under police escort, almost 1,000 women blocked the mill gates.

The police contingent consisted of four vans, carrying policemen with lathis, armed policemen, and policemen with tear gas. It was led by the acting police commissioner. According to the *Bombay Chronicle*, 'the eyes of the whole armed camp were riveted' on the women pickets; the police commissioner told Dange and the communist trade union organiser Nimbkar that if the women did not move away from the gates, they would use tear gas; the women did not move, tension mounted, and in the hope of breaking the deadlock, the commissioner agreed to wait till Dange had phoned the Chief Minister.[65] The phone call stretched into fifteen minutes; the police began preparing to clear a path to the mill gates. Meanwhile, Ushabai Dange and the women pickets surrounded the lorries and pleaded with the blacklegs not to break the strike.[66] When Dange returned to report his failure to wrest any concession from the Chief Minister, 'suddenly the workers in the two lorries said they had decided not to go into the mill'.[67] According to Parvatibai Bhor,

> The women all joined hands and blocked the lorry. Ushatai said to the blacklegs, 'If you want to come with the lorries over the bodies of these women, you can do it.' But after hearing these words they were ashamed, their manhood rose again. They said they would not let any harm come to the women. The police were also surprised. They were afraid they would have to fire on us.[68]

The Phoenix strike lasted for seven months, during which 'the women workers had practically taken possession of the mill entrance and nobody was allowed to enter'.[69] The action was called a *satyagraha*.

Dwarkadas described the Phoenix strike as the first large-scale stay-in strike in the Bombay cotton textile industry. Prior to this, weavers in the Khatau Makanji Mill had gone on a four-day occupation of the mill in November 1936, and had won their demands.[70] The February 1938 strike of women winders at the Suryodaya Mill also began as a stay-in strike.[71] The Bombay government first began to examine 'stay-in strikes' in 1936, when the governor asked the Home Department to investigate the legal position. In its note on the subject, the Home Department commented: 'the stay-in strikes in France seem to have had their repercussions in French territory in India'; in July the workers of three textile mills in Pondicherry occupied their mills, keeping the officials within the compound. Whether a chain of influence can be traced from France through Pondicherry to Bombay is not clear, though it seems likely that the Bombay trade unionists would have known of the Pondicherry strikes.

While the legal position, said the Home Department, was that an occupation constituted criminal trespass, and could also be declared an unlawful assembly, the 'practical difficulties' of ejecting hundreds of strikers required thorough examination, especially 'in view of the possibility of the communists adopting similar tactics in the event of future strikes in this Presidency'. The commissioner of police suggested that tear gas would be the best method as it would 'obviate damage to machinery which would otherwise be liable to suffer were some other methods of dispersing the assembly by force employed'.[72] Yet when these fears were realised by the strike, the police were loath to use tear gas – so very much so that the occupation continued for seven months.

Perhaps the one thing the Home Department had not anticipated was that the chief culprits would be women. The police were in general extremely reluctant to deal harshly with women workers (during the Phoenix strike, for example, the government finally pleaded with the union not to deploy women pickets),[73] and while there is mention of tear gas being used against nationalist women's demonstrations, it seems never to have been used against women workers.

Three years after these discussions within the Home Department, the police were again enquiring of them what tactic to use in the stay-in strike at the Simplex Mills, where 200 women winders had been sitting in the compound for two days without food. The women did not dare to leave the compound for fear that they would not be readmitted (retrenchment again), while the management would not allow food to be brought in for them. The government was not sure what legal action to take against the strikers, because the complaint against them had been lodged by mill clerks, not the management. Notes were still being exchanged by 20 November,[74] though the strike had been called off two days earlier and the dispute referred to the Industrial Court for arbitration.[75] The women had gone without food for five days. They occupied the mill again in December, this time for two and a half days.[76] In October 1947 a stay-in strike of eighty women winders lasted for eleven days.[77]

The use of the term *satyagraha* to describe the methods adopted by the Phoenix strike was not new. Gandhian tactics were adopted by mill workers and labour leaders, albeit to a limited extent and often in mock seriousness. During a strike in 1934, for example, when the mills had as usual withheld wages, as they so often did, 100 women and 400 men decided to offer *satyagraha* at the New Great Eastern and New City of Bombay mills. At the New Great Eastern, they sat in non-violent protest until the afternoon, when the management put up a notice saying that wages would be paid two days later, but these would consist of only nine days' pay as the management 'was entitled to forfeit fourteen days' wages because no notice of suspension of work had been given'. The *satyagrahis* arose in wrath, tore up the notice, shouted 'the usual Red Flag slogans', and after the police drove them to the opposite side of the road some of the men started to stone the police. The non-violent protest ended in a police firing in which six workers were injured, three of them severely.[78] In the same strike, when the commissioner of police imposed a ban on workers' meetings, strikers 'defied the ban' by hoisting the red flag on Golanji Hill in Parel (just as nationalists hoisted the Indian tricolour).[79] And in the general strike called by the Red Flag Union in 1940 against the high cost of living, 400 women mill workers went on hunger fast at Kamgar Maidan in Parel, partly in appeal to workers who had not joined the strike, and partly to protest against police action. Apparently 1,000 women had volunteered for the fast.[80]

Commenting on the 1940 strike, the *Times of India* remarked that 'a new feature' was the mass picketing by women, but this was, in fact, a repeat of the tactic used during the Phoenix strike. The new feature, perhaps, was the wide arrests of women pickets. Thirty-six were arrested by the early afternoon of 4 March; at the E.D. Sassoon Turkey Red Dye Works, when the police arrested one woman, the others with her demanded that either they should all be arrested or none should be. When the police refused to oblige, they followed the van. By the end of the day over 120 women had been arrested.[81] The police responded by banning assemblies, the union cut down the number of pickets to no more than two; but groups of women sat nearby 'apparently to take the place of pickets who might be arrested'.[82] Arrests continued to the end of March.[83]

The first instance of physical intimidation of women strike-breakers by women strikers was reported during the 1940 strike. The weapon was chilli powder:

> A case of intimidation reported to the police concerns five women strikers who are alleged to have dragged two loyal workers out of the Kohinoor Mill into a private compound and attacked them with chilly powder. A considerable quantity of chilly powder was stated to have been sprinkled over the heads of the loyal women and some of it blown into their eyes. The incident occurred at the foot of the Tilak Bridge, Dadar.[84]

Systematic accusations of strikers intimidating non-strikers first began to be made during the 1919–29 strikes, when both mill owners and the government began to collect, and on occasion publish, complaints. In June 1929, the Mill Owners' Association published forty-eight complaints, eight of which concerned women workers, and four the wives of male workers. The one group of complaints was that union volunteers went to their *chawls* and threatened their wives with dire consequences if they were not able to prevent their husbands from going to work; of the other group, four were complaints by *naikins*, one of whom was alleged to have been 'hammered by some Agris' on her way to the mills.[85] Of the 396 complaints registered by the police, only twenty-seven were filed by women.[86]

Union and strike mobilisation within the *chawls* was an especially sore point. Both mill owners and government saw the forming of *chawl* committees and their mobilising for strikes within *chawl* areas as a form of intimidation. Yet *chawl* areas had always been the places where industrial grievances were vented and strikes brewed. Partly because most mill workers lived in densely populated areas within a short walking distance of their mills, relationships between the workplace and the home were close, and workplace struggles extended naturally into the neighbourhood, while the latter could often serve as the arena in which battles were fought out.[87] Kin, caste and regional networks of recruitment were also networks for settlements.

Available literature tells us little about the part women played in this. It was only during the 1940 strike that newpapers reported that women workers had begun to go from house to house in tenement areas to persuade textile workers to remain on, or join in, the strike.[88] But women's involvement in neighbourhood mobilisation appears to have begun much earlier, and there is, indeed, some evidence to show that it was through women that the labour contractors' hold over workers was weakened. In the 1928 strike, when the Girni Kamgar Union turned its attention to women, it found them invaluable in developing *chawl* bases: 'Women volunteered their detailed knowledge of life in the *chawls*, acting as guides for the leaders and helping them to build up a network of communications parallel to that of the jobbers.'[89]

It seems, too, that the traditional authority of the old woman over the younger one could be invoked in favour of a strike:

> I remember after the Second World War started we, on the 2nd of October, called an anti-war strike – a political strike. In that, the whole of Bombay was closed. I was going to some place – going through the mill area – I remember a young woman was going. An old woman saw this young lady going and asked her where she was going. She said: I am going to the factory. Why? Have you got any lover there? Or something like that, the old woman shouted. So women are terrible in these things. It is very difficult to stand against them.[90]

CONCLUSION

Women textile workers organised against their employment in a number of ways, from lightning strikes to joint representation to strike mobilisation and neighbourhood organisation. They were among the first to protest the rationalisation measures whose chief thrust was the massive retrenchment of women.

This chapter deals chiefly with the forms of women textile workers' protests, and one of the more interesting findings of the literature is the specific gender-based role which they played in the textile workers' movement. Their use to shame strike-breakers is especially notable, though one question which arises is how far this role was constructed by the mainly Brahmin 'western-educated' trade unionists. The very sparse literature available on pre-union strikes indicates that there might have been some pattern of equal representation by sex in the early phases of the textile workers' movement. As union representation of workers' demands developed, women do not seem to have been selected as representatives. At the same time, the communists, who led the trade union movement, focused on women's militancy in specific ways.

Almost every labour leader of the time commented on women's militancy. 'Women workers constituted a strong fighting force of the union,' said Chitnis of the Girni Kamgar Union, 'the thing about them was that once they decided on anything they were united.'[91] Sripat Ghone, of the same union, did not agree that women were active in the union, but said they were extremely militant: 'women workers did not take much interest in union work but were always at the forefront of the struggle'. Summing up, Paul d'Souza said, 'all said and done, women led struggles'.[92]

Unfortunately, material for the pre-1920s period is scanty and so only tentative generalisations about the shifts in women's forms of protest are possible. The late 1920s appear to mark a recognition of 'women's activism' as a distinct category, and the years to follow saw the structuring and expansion of this category. Paralleling this, attitudes towards women's militancy appear to have undergone considerable change between 1928 and 1947. During the early part of the period, what comments there were were cautiously sympathetic, but by the 1930s a distinct note of disapproval crept in. (Even so, the contrast with statements about feminine activism in the west is marked, but this is another subject.)

Interestingly, even in disapproval they denied the agency of women. The mill owners, for example, believed that the communists made special efforts 'to organise the women workers into a militant body and to gain their confidence by false propaganda'; the Phoenix and Simplex strikes gave the communists the opportunity, they said, to create the impression that the Red Flag was 'their only real friend'.[93] It was as if any claim by communists to represent women was a particularly unkind cut, even if this was an interest-based claim to working-class representation, and this might have been because of the very strong influence of the social reform movement upon the formation of the Indian middle class and elite – and a central point of that movement was the position of women.

Yet relatively few women were unionised at the time. In 1938–9, only 3.3 per cent of trade union members were women; in 1939–40 this figure had gone up to 10.1 per cent, but fell again to 5.9 per cent in 1940–1.[94] Despite the fact that women trade unionists had appeared by 1928–9, women's representation in the trade union movement was minimal. All three of the women trade unionists who appeared were middle or upper class; two of them, Ushabai Dange and Mrs Nimbkar (even her first name does not appear in reports), were communists and married to communist trade union leaders of the Bombay textile workers' movement. The third, Maniben Kara, was a socialist. A fourth, who rarely appears in documents of the time but who has been referred to here and whose autobiography was published in Marathi, was Parvatibai Bhor, a lower-caste and 'low-class' woman. Illuminatingly, if the other three were middle or upper class, their writings do not appear to have been circulated, and Ushabai's autobiography, published in Hindi, is even more difficult to find than Parvatibai's.

Representation was, in fact, a sensitive issue in the textile workers' movement. While the first salvoes against 'outsiders' were fired at moderate trade unionists, there was tension over the issue in the communist Girni Kamgar Union from its inception. 'Every labourer must be able to make speeches,' said Kasle, a textile worker and trade unionist, in 1928;[95] he and Alwe (see p. 62) in fact dissociated themselves from the 'outsiders' Dange, Nimbkar *et al.* when they were prosecuted by the British for conspiracy against the state, an act Dange commented on bitterly.[96] Whatever the rights or wrongs of the argument, while several male workers did become union leaders, no women workers appear to have done so – or for that matter, to have addressed a public gathering. Almost all the trade unionists of the time confessed that they had not attempted to enrol women as union members, despite the presence of Mrs Dange, though none of them knew why, and all were puzzled at this unaccountable lapse, given the militancy of women. Perhaps S.A. Dange came closest to providing an answer when he said, 'Well, men don't understand women.'[97]

GLOSSARY

babu	clerk
badli labour	substitute labour casually employed
bidi-making	local cigarette making
chawls	tenement housing
dalit	the 'untouchable' caste
dharna	sit-down protest
mukaddams	labour contractors
naikin	woman supervisor
pali system	rotating work system
satyagraha	battle for truth; also a Gandhian term for a non-violent form of resistance
tarwalla	piecer

NOTES

1 Summarised from M.D. Morris, *The Emergence of an Industrial Labour Force in India: A Study of the Bombay Cotton Mills, 1854–1947*, Oxford University Press, Bombay, 1965, pp. 27–31.
2 Richard Newman, *Workers and Unions in Bombay, 1918–1929*, Canberra, Australian National University Monographs on South Asia, No. 6, 1981, pp. 4–5, 152–3.
3 See *Census of India*, 1921, Vol. IX, Part II, p. xiv; Kanji Dwarkadas, *Forty-five Years with Labour*, Asia Publishing House, New Delhi, pp. 98–100; R. G. Gokhale, *The Bombay Cotton Mill Worker*, The Mill Owners' Association, Bombay, 1958, p. 17.
4 A.R. Burnett-Hurst, *Labour and Housing in Bombay*, P.S. King & Son Ltd, London, 1925, p. 10.
5 Morris, *The Emergence of an Industrial Labour Force in India*, pp. 46–51.
6 Labour Office, Government of Bombay, *Report on an Enquiry into Wages and Hours of Labour in the Cotton Mill Industry, 1926*, Government Central Press, Bombay, 1930.
7 Mira Savara, 'Changing Trends in Women's Employment: A Case Study of the Textile Industry in Bombay', Ph.D. Thesis, Department of Sociology, Bombay University, 1981, p. 141.
8 Morris, *The Emergence of an Industrial Labour Force in India*, Chapters VIII, IX, X, also 'Summary and Conclusions', pp. 129–98.
9 N.M. Joshi papers, Nehru Memorial Museum and Library, Delhi, p. 748.
10 Labour Office, Government of Bombay, *Report of an Enquiry into Wages and Hours of Labour in the Cotton Mill Industry, May 1921*, Government Central Press, Bombay, 1923, pp. 35–6; *Report of the Textile Labour Inquiry Committee, 1937–38, Vol. I, Interim Report*, Government Central Press, Bombay, 1940, pp. 23–30.
11 Radha Kumar, 'City Lives: Women Workers in the Bombay Textile Industry, 1911–47', Ph.D. Thesis, Department of History, Jawaharlal Nehru University, Delhi, 1992, pp. 240–8.
12 Mill Owners' Association, *Annual Report*, Bombay, 1914, p. viii.
13 *Bombay Chronicle*, 15 October 1914.
14 S.A. Dange, interview.
15 *Labour Gazette*, October 1923.
16 *Labour Gazette*, April 1926.
17 *Labour Gazette*, February 1928.
18 *Labour Gazette*, July 1928.
19 *Times of India*, 4 January 1928.
20 *Times of India*, 16 June 1928.

21 *Times of India*, 6 June 1928.
22 *Times of India*, 28 July 1928.
23 *Times of India*, 29 June 1928.
24 *Times of India*, 29 June 1928.
25 *Times of India*, 2 July 1928.
26 Newman, *Workers and Unions in Bombay, 1918–1929*, p. 200.
27 *Times of India*, 3 August 1928.
28 *Times of India*, 3 July 1928.
29 Actually Mang Garudi, a sub-caste of the Mangs.
30 *Times of India*, 5 July 1928.
31 *Times of India*, 6 July 1928.
32 *Times of India*, 22 June 1928.
33 *Times of India*, 1 August 1928 and 3 August 1928.
34 *Times of India*, 18 July 1928.
35 Janabai Prabhu and Bikibai Krishna Guldekar, interviews.
36 G.V. Chitnis, interview.
37 S.A. Dange, interview.
38 S.V. Ghate, interviewed by Dr A.K. Gupta and Dr Hari Dev Sharma, Nehru Memorial Museum and Library, Oral History Transcript No. 24, p. 49.
39 *Times of India*, 1 September 1928.
40 At the end of May 1928, for example, a jobber at the Pearl Mill called Krishna Raju, and a fireman at the Century Mill called Dukki Padavat were caught by pickets on their way to the mills, their faces were blackened and they were brought to the meeting at Nagu Sayaji's Wadi and made to apologise (*Times of India*, 1 June 1928).
41 *Times of India*, 30 April 1930.
42 G.V. Chitnis, interview.
43 Police report of a meeting at Bakarayacha Udda, Chamelibag, on 2 September 1928 (Home Department, Special file no. 543 (10) C, Pt. A, p. 116).
44 They attempted to do the same in the 1934 strike, announcing the opening of the mill with a grand *puja* (religious rite). The managers, strikers and trade union leaders all believed that if they succeeded in reopening the mill with a full complement (5,000) then the strike would break (*Times of India*, 17 May 1934).
45 Home Department, Special file no. 543 (10) C, Pt. A of 1928, pp. 158–62.
46 *Times of India*, 23 August 1928.
47 *Times of India*, 25 April 1929.
48 *Times of India*, 29 April 1929.
49 *Indian Textile Journal*, XXXIX, June 1929, pp. 310–11.
50 Newman, *Workers and Unions in Bombay, 1918–1929*, pp. 239–44.
51 *Labour Gazette*, October 1923, November 1923, September 1925.
52 *Times of India*, 2 August 1929.
53 *Labour Gazette*, August 1933.
54 *Labour Gazette*, April 1934.
55 *Labour Gazette*, November 1935.
56 *Labour Gazette*, September 1939.
57 *Labour Gazette*, October 1936.
58 Dwarkadas, *Forty-five Years with Labour*, p. 99.
59 According to the Red Flag Union, they wanted one woman reeler to attend one frame instead of the prevailing system of two reelers per frame; siders were asked to attend two sides for counts below 30s, even 24s and 20s, where the yarn breakages were considerable; and in the waste-picking department, having dispensed with *bigarees* (waste removers) they wanted women waste pickers to take on this job as well (*Bombay Chronicle*, 21 April 1939).

60 G.V. Chitnis, interview; Parvatibai Bhor, *Eka Ranaraginichi Hakikat, as told to Padmakar Chitale*, Lokavadmaya Griha, Bombay, 1977, p. 62.
61 *Bombay Chronicle*, 13 April 1939.
62 *Bombay Chronicle*, 21 April 1939.
63 There had been no alteration in conditions of work, he said, except in the case of six women reelers for three days; therefore the strike was unjustified and the workers were not entitled to any compensation for the days the mill was closed (*Bombay Chronicle*, 11 May 1939).
64 *Bombay Chronicle*, 13 May 1939.
65 *Bombay Chronicle*, 19 May 1939.
66 Dwarkadas, *Forty-five Years with Labour*, pp. 55–6.
67 *Bombay Chronicle*, 19 May 1939.
68 Bhor, *Eka Ranaraginichi Hakikat*, p. 64.
69 Employers' Federation of India, *General Strike in the Bombay Cotton Mills: Lessons for Indian Industries*, Patel House, Bombay, 1940, p. 8.
70 Commissioner of Police, Review No. 82, Home Department, Special file no. 543 (13) B (2) of 1938.
71 Commissioner of Police, Review No. 85.
72 Home Department, Special, UORNo. SD-3193 of 9 October 1936, file no. 550 (19) of 1936, p. 2.
73 *Bombay Chronicle*, 23 May 1939.
74 Home Department, Special, UORNo. 247, 20 November 1939, file no. 550 (19) of 1936, p. 31.
75 *Labour Gazette*, November 1939.
76 Employers Federation of India, *General Strike in the Bombay Cotton Mills*, p. 8.
77 Mill Owners' Association, *Annual Review of the Labour Situation*, Bombay, 1948, p. 5.
78 *Times of India*, 15 May 1934.
79 *Times of India*, 28 May 1934.
80 *Times of India*, 27 March 1940.
81 *Times of India*, 5 March 1940.
82 *Times of India*, 7 March 1940.
83 *Times of India*, 28 May 1940.
84 *Times of India*, 8 April 1940.
85 *Indian Textile Journal*, Vol. XXXIX, June 1929, pp. 134–5.
86 Home Department, Special file no. 343 (10) E (BB), pp. 70–111, 123–9, 131–3, 135–45.
87 Raj Narain Chandavarkar, 'Workers' Politics in Bombay between the Wars', in C. J. Baker, G. Johnson and A. Seal (eds) *Power, Profit and Politics*, Cambridge University Press, 1981, pp. 603–47.
88 *Times of India*, 26 March 1940.
89 Newman, *Workers and Unions in Bombay, 1918–1929*, pp. 201–2.
90 S.V. Ghate, interview.
91 G.V. Chitnis, interview.
92 Sripat Ghone and Paul d'Souza, interviews.
93 Employers Federation of India, *General Strike in Bombay Cotton Mills*, pp. 8f.
94 Mira Savara, 'Changing Trends in Women's Employment: A Case Study of the Textile Industry in Bombay', unpublished Ph.D. thesis, 1981, Bombay University.
95 Home Department, Special file no. 543 (10) C Pt. A of 1928, p. 147.
96 S.A. Dange, *Selected Writings*, Lok Vangmaya Griha Ltd, Bombay, 1979, Vol. 3, pp. 272–4.
97 S.A. Dange, interview.

Chapter 3

The conditions and organisational activities of women in Free Trade Zones

Malaysia, Philippines and Sri Lanka, 1970–1990

Kumudhini Rosa

ABSTRACT

The creation of Free Trade Zones from the 1970s has generated much analysis and controversy, particularly in relation to their consequences for the young women workers they have tended to employ.

They presented problems for earlier forms of labour militancy and have sometimes appeared to be well-nigh impossible to organise.

Kumudhini Rosa's approach breaks new ground by looking at a whole range of organisational forms which are not part of established structures and which link home and workplace.

She examines the process by which a new labour force has been created and directs attention to a crucial question which has been neglected. How does this new kind of Third World woman see herself and her situation?

The Free Trade Zones do not constitute static forms of production. Shifts in investment policies means they are in the process of changing. Understanding of their economic development is inseparably linked to comprehending the nature of the social awareness they have brought to a new generation of workers, many of whom came straight from a rural setting.

INTRODUCTION

The entry of large numbers of women into the Free Trade Zones' (FTZs) labour force, as a result of the restructuring of capital in the late 1960s, is of considerable economic and social significance. Owing to the setting up of FTZs, a new stratum was added to the already existing and often organised workforce. This new layer of workers is mainly composed of young women, of whom a significant percentage are single and have limited experience in waged employment. In many places the women came from rural areas and lived either within the FTZ compounds or within close access to the zone, in boarding houses. Within these 'industrial zones' various firms are encouraged to set up factories which produce for the world market. Special authorities are granted the responsibility for all negotiations with foreign and local investors, control of labour relations, implementation of labour laws and in some cases recruitment of labour.

The peculiar circumstances of employment in the FTZs have both contributed towards organisation among these workers and created certain obstacles. As these workers are largely women who frequently live together and are employed in very similar types of jobs, their situation does provide opportunities for organising. However, workers employed in FTZs are often in waged employment for the first time in their lives, and confront special constraints in organising themselves. The anxiety of the state to maintain a 'stable' environment for investors often ends with state control over labour and non-implementation of labour laws. This makes these workers more vulnerable than workers outside FTZs who have experienced unionised employment in factories.

A description of the arrival for work in the Philippines conveys how exposed workers are to coercion.

> Three and a half hours out from Manila, the bus to Mariveles, the small town near the zone, has to slow down. Soldiers dressed in combat fatigues and well armed with automatic weapons cast their eyes over the bus before waving it through. Notices announce that you are entering a special security area. . . . The main industrial complex is surrounded by high walls and wire fences. No one can enter without a pass. All workers have to carry identity cards and have to queue while they are security checked in and out . . . this industrial zone is governed by its own armed police force with its own intelligence service and network of spies.[1]

The extent to which employment in the FTZs in developing countries is of benefit to women remains a controversial subject. Some recent contributions have stressed the gains for women in terms of greater freedom and status arising from earning a wage. Others have more reservations, pointing to the precarious nature of such employment in many countries, and to the fact that it is generally only available to women for a short period of their lives.[2] Diane Elson and Ruth Pearson have suggested that the gains for women through employment in FTZs should not only be assessed in terms of wages but also take account of opportunities to acquire a sense of responsibility, co-operation and solidarity with other women. Becoming organised helps to develop the capacity for self-determination. While the economic gains may be limited and temporary owing to the mobility of capital, organisation can enable women to develop self-confidence and a new social identity and to acquire new skills.[3]

A variety of forms of resistance and organisation have indeed emerged among women workers employed in FTZs. However, these have not been systematically examined globally. By focusing on just three countries – Malaysia, the Philippines (Bataan FTZ) and Sri Lanka (Katunayake FTZ), it is possible to see that various organisations, including trade unions and women's organisations, have faced considerable risks to reach FTZ workers. Women workers themselves have spontaneously developed both overt and covert forms of defence and members of these organisations have often been arrested.

In all three situations the legal framework and the direct intervention of the state have influenced the methods of resistance. The organisational forms adopted are not static but continually evolving in response to economic and social change. Their development is based on an accumulating experience which has been gained through past efforts, failures as well as successes in the FTZs.

Access to material about the organising and resistance of women workers in FTZs remains difficult. While some accounts of significant strikes and campaigns find their way into print, often in international labour bulletins and women's newsletters, much of the history made daily on the factory floor remains completely undocumented. My personal involvement and practically based knowledge of the Sri Lankan case has enabled me to draw on oral sources for the period 1978–88 which give greater depth to the assessment of the changes in consciousness which occurred among the women. In Sri Lanka the new wave of the women's movement arose in the late 1970s, at the same time as the FTZ. Many groups began initiating discussions in the villages about the effects FTZs would have upon the community. I was involved with other women in setting up a women's centre to support the workers. We provided legal and medical assistance, reading and library facilities, a place for socialising and meeting. Through this experience, I was able to develop an understanding of the organisational capacities of the women workers, and learn how they interpret their own experiences and life stories. Since then I have co-ordinated research in the FTZs in Sri Lanka, worked closely with such organisations as Women Working Worldwide and edited *Drops of Sweat*, a newsletter which provided information of women workers' struggles globally.

This account of organisation and resistance among women workers employed in the FTZs in the 1970s and 1980s is consequently based on a mix of participatory learning and research which draws on international labour networks as well as official sources.

DEVELOPMENT OF THE FREE TRADE ZONES

The FTZs in Malaysia were established under the FTZ Act of 1971, and function subject to the FTZ (Manufacturing Regulations), 1972. By the year 1980, eight FTZs were functioning all over the country; three more were set up after 1980.

Diverse items are produced within the Malaysian zones, 90 per cent of which must be exported. The product range encompasses electronic components, metal parts, machinery equipment, optical and photographic goods, rubber products and textile goods.

Two types of industry coexist within the Malaysian FTZs: first, local subsidiaries of multinational corporations (MNCs) carrying out simple assembly operations (employing 'unskilled' female labour) where raw materials are imported and assembled to be re-exported to the parent company and, second, integrated industries producing completed goods. Both types are in most cases MNCs or 100 per cent foreign-owned companies. (For instance among the firms

Table 3.1 Structure of employment in the FTZs in Malaysia as at 1979 (in percentages)

Electronic/electrical machinery	74.5
Textiles and garments	14.2
Instruments and optical products	3.7
Rubber products	2.6
Metal products and machinery	5.8
Total	100.0

Source: *Economic and Social Effects of Multinational Enterprises in Export Processing Zones*, International Labour Organisation and United Nations Centre on Transnational Corporations, ILO, Geneva, 1988, pp. 38–9 (extract from Table 5)

in the Batu Berandam FTZ were subsidiaries of National Semi Conductors – NS Electronics – a US multinational and the German company Siemens, employing about 3,000 workers each.) A number of joint-venture enterprises operating with Malaysian companies also exist. The electronics industry, established in 1971 with about 15,000 workers, was to comprise nearly 75 per cent of the investments by 1979 (see Table 3.1). By 1986, there were over 85,000 workers employed in over seventy electronics companies, owned mainly by US or Japanese MNCs.

In the Philippines (Bataan FTZ) the first FTZ was set up in 1972. Two more followed, one in Mectan Island and the second in Baguio. The FTZ was set up under Marcos's martial law. Although very attractive incentives were offered to investors, the study conducted by the Asia Partnership for Human Development (APHD) in 1985 shows that the results have not been as expected. By 1983, only forty-one firms were operating in the Bataan FTZ. Eight were wholly Filipino owned; twelve were joint ventures with Japanese or other foreign firms. Hong Kong, South Korean, Taiwanese investors began to show interest when they found they could benefit from Philippine export quotas granting access to protected markets, and take advantage of the cheaper labour offered compared to that of the Newly Industrialised Countries (NICs).[4] In 1983, seventeen of these forty-one firms were engaged in light manufacture (footwear, plastic and rubber products, packing materials, toys, food, etc.), thirteen firms were in garments, five in electronics, four in heavy fabrication and two in ship building.[5]

By the second half of the 1980s there were two FTZs operating in Sri Lanka. The first was set up in 1979 in Katunayake (15 miles north of Colombo) adjoining the only international airport, and the second in Biyagama (14 miles from Colombo). The Biyagama FTZ was established in December 1985. A third was to be opened in Koggala, with access to the natural harbour of Galle, in the south of Sri Lanka in 1992. By December 1990 the Greater Colombo Economic Commission (GCEC), the special authority responsible for FTZs in Sri Lanka, had approved 337 projects; 120 of these were in commercial operations, 65 being in the Katunayake FTZ and 23 in the Biyagama FTZ. The rest were in the Greater Colombo Economic Area, thus under the authority of the GCEC. In 1988, 73 per cent of the total number of employees (around 54,000) were engaged in the textile, clothing and leather products sector, the dominant production sector for the FTZs in Sri Lanka. Other products included porcelain figurines, fishing gear, rubber and metallic goods. Electronics firms were very few. Investors came mainly from Taiwan, Hong Kong, South Korea and Singapore, though US, West German, British and Belgian firms also operated both as single and joint holders with Sri Lankan firms. As in the Philippines, the main reason for foreign firms to move into the Sri Lankan FTZs is the low labour costs and the access to Sri Lanka's unfulfilled Multi Fibre Agreement (MFA) quotas for garment exports to developed countries.

The dominance of electronics firms in the Malaysian FTZs makes it a somewhat different case from the Philippines and Sri Lanka where firms producing garments and clothing dominate. This singularity is also evident in the extent of the involvement of MNCs in the manufacturing sector and the high participation of industrialised countries in the Malaysian FTZs (see Table 3.2). Workers employed within electronics enjoy more benefits, and work in larger, more comfortable work environments; this distinguishes their expectations and aspirations from those of workers in the garment and textile manufacturing sectors who are employed mainly by Asian investors.

THE COMPOSITION AND BACKGROUND OF THE WORKFORCE

The International Labour Organisation's statistics for 1986 indicated that 1.3 million workers were employed in FTZs in the developing countries.[6] This figure was expected to rise significantly during the next years. In the Philippines, Malaysia and Sri Lanka 85–90 per cent of the workers employed are women. They are between the ages of 17–29 years (see Table 3.3) and the majority came from the rural areas or neighbouring villages into the cities to be employed in the FTZs.

The APHD survey conducted in the Batu Berendam Zone in the state of Malacca in Malaysia, among 287 workers, revealed that the parents of the workers (83 per cent from the rural areas) were poor. Fifty-nine per cent of their fathers were agricultural labourers, working as rubber tappers or in palm oil land. Others were either small-scale businessmen, self-employed or low-grade government employees. The mothers were either housewives or themselves rubber tappers. As a result of conscious government policy in the wake of the 1969 communal disturbances, the vast majority of the workers were Malay. The APHD survey showed that by the mid-1980s, 92.3 per cent of the workers were Malay, 5.6 per cent Indian, 1.4 per cent Chinese and 0.7 per cent others. Of the young women, 95.8 per cent were not married and their level of education was low compared to their colleagues in both Sri Lanka and the Philippines.

The workers coming within the APHD survey were paid a wage rate of M$4.70–M$5.50 for eight hours of work (M$5.00 = US$2.10 at the time of the survey); 83 per cent of the workers received less than M$300.00 per month (below the official poverty line). It must be noted that Malacca was, in

Table 3.2 Malaysian American Electronics Industry (MAEI) member companies (as of May 1989)

Company	*Plant location*	*Year established*
Advanced Micro Devices Sdn Bhd*	Penang	1972
Applied Magnetics (M) Sdn Bhd	Penang	1988
Harris Semiconductor (M) Sdn Bhd†	Kuala Lumpur	1974
Hewlett Packard (M) Sdn Bhd	Penang	1972
Imprimis Technology-Components (M) Sdn Bhd	Penang	1988
Integrated Device Technologies	Penang	1988
Intel (M) Sdn Bhd	Penang	1972
Maxtor Singapore Ltd	Penang	1988
Monsanto Electronics	Kuala Lumpur	1972
Motorola Malaysia	Kuala Lumpur	1972
Motorola (M) Sdn Bhd	Penang	1974
Motorola Semiconductor (M) Sdn Bhd	Seremban	1979
National Semiconductor Sdn Bhd	Penang	1971
National Semiconductor Sdn Bhd	Malacca	1972
Quality Technologies Optoelectronics (M) Sdn Bhd	Kuala Lumpur	1979
RCA Sdn Bhd†	Kualu Lumpur	1973
Siemans Litronix (M) Sdn Bhd	Penang	1972
Texas Instruments (M) Sdn Bhd	Kualu Lumpur	1972
Western Digital (M) Sdn Bhd	Kuala Lumpur	1973

Notes: * AMD merged with Monolithic Memories Incorporated in early 1988.
† Harris acquired RCA in late 1988 and is in the process of merging Harris's and RCA's operations in Kuala Lumpur's Ulu Klang FTZ.

Source: *Malaysian American Electronics Industry White Paper 1988–89*, Kuala Lumpur, 1989, p. 15

comparison to Penang, Selangar, a low labour location for investors. By the early 1980s Malaysia's zones were employing around 70,000 workers. In January 1985 in the Penang Zone alone, where 259 firms were in operation, 62,400 workers were employed. Thirty-one firms operating in the Bayan Lepas Zone in 1982 employed 30,000 workers. In Malacca, in the Batu Berandam FTZ, the workforce grew from 3,100 workers (1970) to 12,500 (1978).[7]

The APHD survey conducted among the workers in the Bataan FTZ in the Philippines demonstrates that 97 per cent of workers were migrant workers, only 3 per cent from Mariveles itself. They too originated from very poor family situations, where the rate of unemployment among the parents was very high. The APHD study established that 33 per cent of the workers' fathers and 80 per cent of their mothers had no (paid) employment.

Table 3.3 Age of Export Processing Zone (EPZ) workers, and share of women in EPZ workforce and in non-EPZ manufacturing industries (early 1980s)

Country or area	*Age of EPZ workers*	*Share of women in EPZ industries (in percentages)*	*Share of women in non-EPZ manufacturing industries (in percentages)*
Malaysia	Average age 21.7 years	85	32.9
Brazil	Average age 21.7 years	48	24.8
Macau	88% below age 29	74	48.1
Mauritius	70% below age 25	79	10.0
Tunisia	70% below age 25	90	48.1
Philippines	88% below age 29	74	48.1
Dominican Republic	83% below age 26	68	17.6
Sri Lanka	83% below age 26	88	17.1

Sources: Case studies prepared in the framework of the ILO-UNCTC project; ILO, *Year Book of Labour Statistics 1986* (Geneva, 1986); P. G. Warr, 'The Jakarta Export Processing Zone: Benefits and Costs', *Bulletin of Indonesia Economic Studies* (Canberra, XIX (3), December 1983, for data on Indonesia; J. E. Shapiro: 'Taiwan – Where the EPZs are Called "Utopia for Business, and 55,000 Young Women Risk their Health for 70 Cents an Hour"', *Multinational Monitor* (Washington, DC) June 1981

The workers were relatively well educated. The survey showed that 59 per cent had completed high school while 19 per cent had enrolled in or completed college studies. The electronics workers were seen to have higher educational qualifications than the garment workers.

Owing to non-standardized wage levels of the workers, the survey found that they received between 25 and 33 pesos a day. A casual worker or those on probation, however, earned 14 pesos a day, and a trainee or apprentice 12 pesos a day (both categories were not entitled to allowances).

Although it was expected that FTZs would generate employment, in the early 1980s the number of jobs in all three FTZs

> dropped from 28,000 in 1980 to 25,610 in 1983. This was only 0.13% of the total Philippine labour force. . . . Retrenchment of workers has continued to be savage. In 1981, the Bataan FTZ employed 20,370 workers. By 1985 that had fallen to an estimated 15,000 workers.[8]

The FTZs in Sri Lanka employed over 60,000 workers by December 1989.9 Two-thirds of the workers came from neighbouring village areas and the others from the rural areas (mainly in the south of the country, 4–5 hours by bus from Colombo).

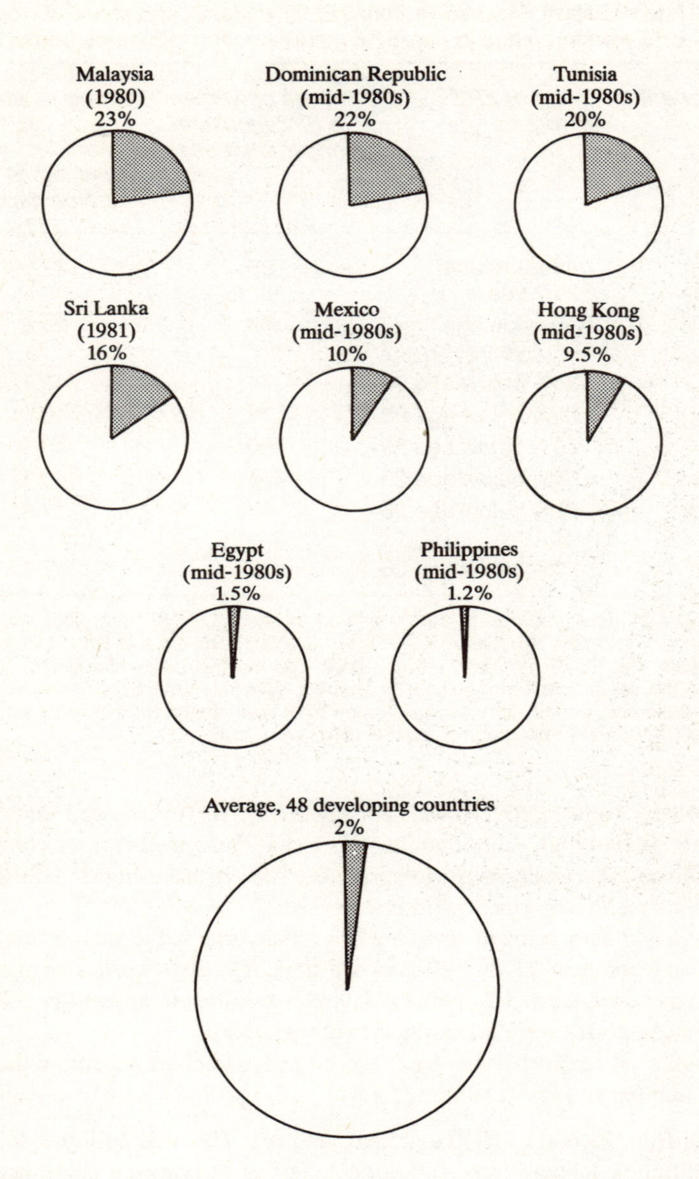

Figure 3.1 Employment in EPZs as a percentage of total manufacturing employment

Source: International Labour Organisation and United Nations Centre on Transnational Corporations, *Economic and Social Effects of Multinational Enterprises in Export Processing Zones*, ILO, Geneva, 1988, pp. 21–2

The women employed were in paid factory employment for the first time in their lives. They were relatively well educated; 8–12 years of schooling meant they were literate and tended to have high aspirations. The Sri Lankan women's publication *Voice of Women* found in 1983 that in this new workforce of women ethnically Sinhala workers predominated, constituting 97 per cent. Some 56 per cent of these workers were Buddhist while 41 per cent were Catholics.[10] Katunayake, Negombo and Andiambalama areas just outside the FTZ are predominantly Catholic which probably explains the heavy Catholic component, for on a countrywide basis Catholics are only 7 per cent of the population. The negligible participation of Tamil and Muslim workers in the FTZ workforce is particularly significant. Workers were recruited through Members of Parliament for the villages from which they came; making the 'correct' political affiliation a necessary precondition for obtaining a job in the FTZ.[11]

The wages are determined by the wages board, calculated on the basis of a minimum rate linked to the items being produced and the 'skill' they possess. A worker could be 'skilled', semi-skilled', 'unskilled' or 'a trainee'. An experienced machine operator is a 'semi-skilled worker' earning Rs. 31.00–Rs. 35.00 per day (equivalent to US$1.10, as at January 1989).

In all three situations the APHD's surveys and my own interviews stressed economic reasons for women's entry into the FTZs. Supporting their families was a recurrent reason given. They frequently saw their employment in the FTZs as a temporary means of livelihood. However, many had developed new aspirations through their employment and the opportunity of earning wages.

In Malaysia, the workers had qualifications higher than those needed for unskilled or semi-skilled work in the electronics firms; they were persistently seeking more permanent jobs, particularly with government departments. The Bataan workers expected that the experience in the zone would provide them 'with a good chance of finding better and more responsible jobs in the future'.[12] Along with expectations of more permanent employment, the Sri Lankan women hoped to have time to educate themselves further. For example, Leela explained,

> I believed that I would have the chance to study further. The day I arrived at the GCEC office, I asked the officer whether I have to work on Saturdays and Sundays. He replied that this would happen only occasionally. But now, by the time Monday or Tuesday arrives, they have already forced us to sign agreeing to work on Saturdays and Sundays. If we do not sign, they remove our time cards and then we have the option to leave the factory and never enter it again.[13]

Not simply working but living conditions have to be taken into account. In Sri Lanka, women who come from the rural areas find boarding facilities within the villages adjoining the FTZ. Prior to the existence of the FTZ these villages were rural and houses were built to accommodate small families. However, after 1978, the villagers found an additional income through renting out rooms to the women workers. The houses were ill equipped for such purposes, being at best inadequate and often appalling.

The women I interviewed usually shared rooms, each room holding up to six or even eight women, who occupy a space where a mat or mattress can be placed with just enough room for a suitcase at its foot. In 1988–9 the average charge for a worker who occupies 'mat space', with shared toilet, bathing and cooking facilities, was Rs. 100–150 per month. One toilet and well were often shared by all the women – sometimes the numbers ranging from forty to fifty workers in one boarding house. The women operated a shift system to ensure that they all had access to facilities which sometimes included food. In 1986–7 the average charge for a worker who occupied 'mat space' and was provided with food was Rs. 400–450 per month.

In the Philippines workers tend to be based in the town of Mariveles (a small town adjoining the FTZ). Although limited housing units are provided within the zone area the workers prefer to live in the town because of access to recreational and community activities, and also they tended to be cheaper than the zone dormitories.

In Malaysia no housing is provided by the FTZ management, compelling workers to find accommodation in private houses. The situation is similar to Sri Lanka where sometimes over twenty women share facilities.

The Asia Partnership for Human Development sums up the social and psychological impact of the gulf between the actual quality of life in the FTZ and the unrealised potential of the new labour force.

> Shift work dictates the total life pattern of the worker. It defines sleeping, recreational and socialising hours as well as working hours . . . tight supervision combined with minimum quotas and fixed work postures have resulted in tremendous stress on the woman worker. The awareness that little or no opportunities exist for her to acquire further skills or pursue her education creates a situation of conflict in her own predicament. She finds herself trapped within the demands of the assembly line and continuing responsibilities to support and help her family.[14]

LABOUR MANAGEMENT

Managements, often supported by the special authorities responsible for the smooth functioning of the FTZs, are extremely concerned to prevent any form of labour organising among the workers. Various tactics have been adopted. These range from social provision, controlled forms of participation and suppression of autonomous organisation. Industrial management is closely linked with political measures of control which include laws, surveillance and intimidation to secure industrial peace.

In Malaysia strict rules were in force in the period 1970–90 for registration of unions and the election of union officers (especially in the electronics industry). The Trade Union Ordinance of 1959 gave the minister the authority to suspend a union on very broad grounds. When suspended, a trade union becomes illegal. The Industrial Relations Act of 1967 provided 'protection of pioneer industries during the initial years of [their] establishment against any unreasonable demands from a trade union'.[15] These pioneer industries included companies operating in the zones, and they were therefore exempt from collective bargaining and strikes. During the 1980s the Malaysian government pursued a corporatist policy which aimed to erode the independence of worker organisations and place restrictions on labour negotiations by forming tripartite councils dominated by government-leaning employers.[16]

Company managers offered non-monetary incentives to boost productivity. Encouragement of workers to compete with each other – by delivering higher production targets – has been a tactic which creates a spirit of rivalry between them. Social activities after work such as dinners, dances, fashion shows, beauty competitions, beauty culture classes, introduce a pseudo culture, and distract workers from their conditions of work, health problems and living situations.

Firms provided comfortable conditions of work (particularly the electronics firms), piped music, transport to work, free or subsidised lunches, medical leave and social activities which, APHD reports, tended to obscure 'the class relationship between the workers and management'.[17]

Although some of these social activities (such as beauty contests) or gifts were offered to workers in the Philippines and in Sri Lanka, the response to and success of such programmes in terms of diverting them from other issues has been different. In fact in Sri Lanka, the women interviewed were openly suspicious of such plans. 'Christmas parties organised with deductions from our salary and not their profits? We prefer to keep our wages'[18] was one worker's indignant response. When FTZ managements have sought to establish welfare societies or sports clubs, the workers have shown little or no interest.

In the Philippines the Marcos regime functioning under martial law from 1972 enacted a succession of oppressive laws to curb labour organising.

> General Order No. 5 enforced an immediate blanket ban on strikes. The labour code of 1974 and a series of presidential decrees institutionalised a new system of labour controls. . . .
>
> Presidential Decree No. 143 abrogated the law reserving Sunday as a day of rest, relieving employers of the obligation to pay overtime rates or special allowances for work on this day. . . . Presidential Decree No. 823 suppressed the right to strike and reinforced General Order No. 5 by prohibiting all organisations, including religious and foreign organisations, from giving direct or indirect support to workers' movements.[19]

Though these labour laws were increasingly challenged by mass actions, they were implemented with intensive repression until the fall of the Marcos regime in 1986.

President Aquino passed the Republic Act 6715, known as the Herrera Bill on 2 March 1989. This was presented as protecting labour by strengthening the constitutional rights of workers to self-organisation, collective bargaining and peaceful concerted activities. However, women workers interviewed by the

Hong Kong-based Committee for Asian Women in 1989 expressed the fear that it was yet another blow to their right to organise. They pointed out that the Herrera Bill narrowed down the acceptable reasons for legal strikes, meaning that demands such as the right to four months' maternity leave, the elimination of sexual harassment or other challenges against discrimination towards women workers were no longer grounds to call a strike.[20]

In Sri Lanka the Greater Colombo Economic Commission has operated somewhat differently. No laws were enacted curbing the rights of workers, although in reality workers possess hardly any rights to organise themselves at all. The GCEC provided a complaints office where workers could lodge complaints. Though there were promises that these would be dealt with, the effectiveness of this office is questionable, as workers said they gained little or no respite through going to it.[21]

Workers' councils have been encouraged in factories, where management with elected workers can sit together to resolve problems. Here again the women complained to me that the elected members do not always express the interest of workers at the shop floor, and the workers' councils remain only on paper. For instance, Leela claimed:

> In our factory there is a workers' council. Representatives have been elected from the different sections to the council. During the first meetings, the workers who talk a lot were given a pay rise. They behave the way the Management wants them to! Our problems go under! . . . After the first couple of meetings the workers' council hardly ever meets.[22]

The GCEC itself intervened in disputes, supposedly to bring about a just settlement – but Leela said:

> The only authority who can intervene directly on our behalf is the GCEC. But when they act so openly on behalf of the management, we realised that we can no longer depend on the GCEC to take our side or the side of justice. Every month at least 25 petitions reach the GCEC from our factory alone. There must be a reason why they refuse to respond to our pleas.[23]

No trade union existed within any Sri Lankan FTZ enterprise even in 1990. Throughout the 1980s trade unions were not encouraged to form branches within the FTZ, although the GCEC brochure states that 'employees have the right to form and join a trade union. Recognition of such unions by employers is not regulated by law'.[24] Nevertheless, the idea that prevailed among the workers was that they are not allowed to form trade unions. This is accentuated by the heavy reprisals meted out to workers for even attempting to organise themselves. Unfortunately the national trade unions themselves have done nothing to change this idea or act directly to intervene within the FTZs.

The level of repression against labour activists or workers who intervened in disputes on the factory floor increased dramatically during the late 1980s. Many workers have disappeared or been killed, merely for speaking up. Fear has been used as a weapon to keep the workers suppressed and intimidated.

On the whole though, the legal restrictions curbing labour organising in the FTZs have been more obvious in Malaysia (particularly in electronics) and the Philippines than in Sri Lanka where restraint has usually been implemented more covertly. Nonetheless, workers have still taken high risks if they stepped out of line.

ORGANISING AND RESISTANCE

Despite the repressive laws and conditions, and enormous pressure to earn wages sufficient to be stretched in order to support their families in the villages, the women workers have developed diverse ways of organising. In some cases these have actually gone beyond the traditional forms prevalent outside the FTZs.

The political and economic contexts of the three countries influence the methods that have been adopted. FTZs became operational at different times: in the Philippines and Malaysia this was in the early 1970s while Sri Lanka came in much later during 1979. In the Philippines the political and labour movement developed an experience and maturity during Marcos's martial law regime, which has little comparison to either Sri Lanka or Malaysia.

The type of industries that the countries have attracted are again different. As Table 3.1 indicates, the electronics firms dominating the scene in Malaysia, and the garment and light industries in the Philippines and Sri Lanka, offered very different expectations to the women workers and therefore have had differing results in terms of their attitudes and aspirations for the future. Moreover the women workers' responses to their situations have played a central role in the surge of new forms of organising and resistance that have developed during the past years in all three countries.

INFORMAL AND SPONANTANEOUS FORMS OF ORGANISING

The first step in the development of organisation is trust. The life in the boarding houses fosters a sense of unity and strength which women share towards each other. The experience of living together as women involves sharing similar problems, and dreaming similar dreams. The women develop a collective responsibility which has strong sustaining powers. Organising the cooking, eating, sleeping, washing, within one place contributes towards building a community life. This community feeling, a sense of belonging, becomes an important support structure, offering them the chance to take the edge off their loneliness at being away from their homes and families.

Women workers have created clandestine means of communication and organisation at work. In situations where union organising is not permitted, eye contact, initiatives of small groups, use of local language in the presence of foreign managements, become very important. Actions that result from these forms of contact are not decisions taken at formal meetings, but spontaneous habits of resistance developed by the women workers among themselves. In the words of a Sri Lankan zone worker, Rohini,

> We have our own ways to organise ourselves. This is very important for us. After a period the workers have got accustomed to these methods. They know how to act given a particular situation . . . for instance if new production targets are introduced. It is at the beginning that it is difficult, but after a while this changes and new and varied methods begin to develop.[25]

Spontaneous strikes in Sri Lanka, 'wild cat strikes' in Malaysia, sympathy strikes in the Philippines are not always preplanned activities of the women workers; rather they emerge out of on-the-spot situations requiring immediate action. Besides strikes, lowering the production target when pressure is placed on them to increase their productivity or helping another worker who is slow to reach her target, have been small but important ways to develop support and survival methods.

FTZ managements have tried often to counter these informal forms of resistance by imposing rules such as 'talking is not allowed' or controlling the use of the toilets, 'one woman at a time'. But these measures have aroused anger which has inspired the women to rebel more.

Organisational support is not confined to the actual place of work. For example in the Sri Lankan context, an important part of the women's daily lives is spent getting to and from work. Transport facilities are not provided for the women who have to report to work at 9.30 p.m. for the night shift. As a result women have to trek through unlit roads which has given a free hand to thieves and rapists. On the road leading from the zone, the women move in groups intent on staying together and getting home soon, occasionally calling out to ensure friends are still there. Although these are not the safest methods, the FTZ women have found them preferable to waiting alone at a bus stop or vainly expecting the authorities to provide them with transport.

These forms of resistance and mutual aid did not suddenly arise when the women found themselves in the FTZ but were adopted from strategies familiar to women already. Living in an authoritarian household, learning to be an obedient student and later a disciplined worker, provides an apprenticeship in concealed ways of rebelling. Externally the women might appear passive to male heads of households or management; in fact they have an extremely sophisticated culture of subversion.

TRADE UNION ORGANISING

In both Malaysia and the Philippines, trade unions have made contact and begun organising women workers in the FTZs. In the Philippines the APHD survey revealed that:

> 41 trade unions were formed in the Bataan FTZ by 1983, covering workers in 89% of all firms operating in the zone . . . 92% of respondents saw unions as important and effective tools in fostering unity . . . a majority of 84% believed strikes were an effective weapon to safeguard their rights.[26]

However, both the national trade union centres – the Malaysian Trade Union Congress (MTUC) and the Trade Union Congress of the Philippines (TUCP) – have come under heavy criticism from other labour unions.

The MTUC was established in 1950 as an umbrella organisation, in the words of the APHD survey, for moderate unions which had not been deregistered for alleged communist infiltrations. It is a union operating on the national level although it cannot itself engage in collective bargaining for its members or enter into industrial action. Its role is defined by Elizabeth Grace in her study of the unionisation of electronics workers, *Shortcircuiting Labour*, as a co-ordinator of private-sector unions 'whom it represents in tripartite, consultative bodies such as the National Labour Advisory Council'.[27]

Grace explains that because the MTUC has been condemned for its 'lack of democracy, bureaucracy, infighting and even corruption',[28] the decision was taken in June 1989 by a number of MTUC affiliates and former affiliates to set up a rival umbrella body, the Malaysian Labour Organisation (MLO).[29]

The lack of participation of women in the principal official positions is another point made against the MTUC: 'except for the chairperson of the women's section who is also a vice-president and sits on a number of committees, only one out of eighteen principal officials is female.'[30] Nonetheless, the MTUC has played a significant role with regard to Malaysia's unorganised electronics workers. It has been responsible for gaining international publicity and securing representations to organisations such as the International Labour Organisation. It has also provided legal advice and a service centre for unorganised workers in Penang. For instance, when women workers at Mostak and Atlas were cut back they were able to benefit from these services, although the relationship between the union and the women was not an easy one.[31] The MTUC was also responsible for the setting up of the National Union of Electronics Industry Workers (NEW).

In March 1978, the Textile/Garment Workers Union of Penang organised the largest textile factory employing 2,300 workers – Woodard Textile Mills, owned by Pen Group of companies. By December 1978, Pentax had also joined the union. This union has held a 'good track record' of being active and genuinely representing the workers in the northern part of the Peninsula of Malaysia where they conduct systematic education programmes for the workers.[32] Several other unions such as the National Union of Rubber Products Employees, working in Bayan Lepas FTZ, and the Electrical Industry Workers Union (1972), organising in the Sungei Way Export Processing Zone in Kuala Lumpur, are also actively involved. They have been constantly under threat from the Registrar of Trade Unions, who has the authority to disqualify a union, rendering it illegal.

The Trade Union Congress of the Philippines (TUCP) has also come under heavy criticism from other labour organisations including the Kilusang Mayo-Uno (KMU) and been called a 'yellow union', recognised and controlled by the government. It was 'granted the right to represent labour in tripartite labour-management-government conferences'.[33] The TUCP is also affiliated to the International Conference of Free Trade Unions (ICFTU) based in Brussels.

The TUCP was heavily resisted by Kilusang Mayo-Uno (KMU or First of May Movement) for supporting government labour strategies such as the claim there should be 'one union – one industry'. The KMU is a labour federation organised in May 1980. By the mid-1980s it claimed a membership of over 500,000 and was advocating a 'genuine, militant and nationalist unionism'.[34] Thirty per cent of its leadership were women, between them constituting a strong representation of women workers.

The labour movement in the FTZ was further strengthened by the formation of AMBALA (Alyansa ng mga mangga awa sa Bataan – Bataan Alliance of Labour Associations), described by the APHD survey as 'an alliance of unions and worker associations, individual workers and friends of Bataan workers . . . working for a genuine, militant and nationalist trade union'.[35]

In September 1988 the Malaysian government suddenly announced that the country's 85,000 electronics workers would be allowed to form or join trade unions. The announcement was sudden because for the previous two decades or so the government had tried to block all attempts by the labour movement to organise the electronics workers. However, two and a half weeks after this September announcement, it declared that the electronics workers would only be permitted to form 'in-house' unions and not join any national or industry-based union.[36]

The MTUC petitioned the International Labour Organisation in January 1989, seeking their intervention in persuading the government of Malaysia to grant the electronics workers their 'basic right of freedom of association without imposing any restrictive conditions'.[37]

In-house unions have both negative and positive implications for women workers and their contradictory aspects need to be addressed. The MTUC in a letter to the International Labour Organisation in Geneva outlines negative features, explaining the 'unsuitability' of in-house unions because

> firstly, the workers by and large are not experienced in trade unionism, secondly, the possibility of victimisation and/or favouritism by the management cannot be ruled out; thirdly, subtle financial and material contribution by the employers to the unions with a view to influencing the leadership is a real possibility; and fourthly, it is bound to create economic imbalance in the industry.[38]

Elizabeth Grace observes another danger:

> in-house unions may lack the independence and strength to rise above management prerogatives. Not only will the union's financial resources be limited and directly dependent on its plant's success, but it will lack independently paid union officials like those in national unions.[39]

However, owing to the fact that in-house unions are small in size and revolve around the workers on the factory floor, it can also be argued that they may afford greater opportunities for direct participation and genuine representation to women workers. In-house unions may provide women workers, who have been historically marginalised by national unions, a chance to be more active around their specific issues. This rank and file mobilisation could force the union leaders to take the needs of all workers more into account.

The declaration by the government in September 1988 had only resulted in the formation of one in-house union by 1990, that of the RCA workers. The RCA plant was established in 1973 in the Ulu Klang FTZ in Kuala Lumpur, employing 3,000 workers. It manufactures integrated circuits, power transistors, thermistors and hybrid devices for export to the US, Europe, Asia, South America, Australia and New Zealand.[40] The RCA workers' union has been in deep dispute with management and several workers have been dismissed.

Given the fact that electronics workers have no other formal channel through which to confront their employers, it is possible that in-house unions could become a place of resistance for workers as well as a transitional step towards being linked to or forming a nationwide union.

During the 1980s women's groups began to emerge within or linked to trade unions. These vary considerably. For example by the late 1980s a women's committee existed in the Malaysian Trade Union Congress. However, Elizabeth Grace's research showed that the women's committee enjoyed few rights and a major disadvantage was 'its lack of autonomy in policy making'.[41] It has also had to operate within an unsupportive environment. Elizabeth Grace noted in 1990, 'my own interviews confirmed that there is also a great deal of condescension within the MTUC leadership towards women'.[42]

The women's committee, however, has had its own funds and was therefore able to design its own educational programme of courses and seminars on the legal rights of women workers, teaching them leadership skills and public speaking. One important achievement of the women's committee has been the setting up of a lodging project in 1982, near Kuala Lumpur's Sungei Way Free Trade Zone. Two houses serve as hostels and have a combined capacity for thirty-two persons. By the late 1980s, the project was providing, according to Elizabeth Grace, 'safe and economical accommodation for electronics workers, as well as educating women about their rights and about unions',[43] thus enabling women earning low rates of pay to find accommodation. Such initiatives can be seen as a positive approach towards making contact with unorganised electronics workers.

A very different kind of women's organisation, the Kilusang ng Mangga-gawang Kababaihan (KMK) in the Philippines, was founded in 1984. A mass organisation for women workers asserting women's specific needs and demands, the KMK was based on the recognition of the need to link with other women's organisations to struggle for 'a genuine democratic society'.[44] Financed partly through membership dues and by other fund-generating projects, by 1987 KMK was claiming a national membership of 20,000 women. It established education and training programmes but put considerable emphasis on encouraging women to organise themselves. Democracy as a process which develops consciousness was a key aspect of its structure. The KMK National Congress consists of representatives of its various chapters and women's committees. In 1988 KMK reported to the Committee for Asian Women that further 'efforts are taken to ensure that decisions made are supported by local chapters'.[45]

The KMK, though formed to campaign for women workers' particular interests, told the Committee for Asian Women that it shared the perspective of the militant labour federation KMU:

> Because of the common vision in promoting the interests of the workers, KMK is now affiliated to the labour union KMU . . . we are not organising to separate women from men but to address problems specific to women. We must demand that women's problems be solved, but we recognise that the overall struggle is one of class.[46]

Trade union organising during the 1980s therefore remained the main medium of organising in the Bataan FTZ in the Philippines, while in Malaysia it played a lower but nevertheless significant role. In both cases women's organisations and committees acted as channels to bring women closer to the trade unions to which they are affiliated.

Women's organisation was not, however, simply a matter of formal structures linked to trade unions. Considerable experience was gained by women workers during the 1980s through participating in the insurgent strike movement in the Philippines. Women had to take considerable risks for the right to organise and consequently freedom in the workplace came to have a deeper and broader meaning. For example, at the Bataan FTZ, a woman worker Lisa described events at the picket line at Lotus Shoes, in June 1985, where the workers were uniting to change their affiliation from the Federation of Free Workers to the KMU.

> At around 9.00 a.m. the military turned up with fire trucks again. They pulled our strike banners down, but we just put more up. Then they rounded us up with guards. We were taken to the zone administration building and interrogated. They took our ID passes and banned us from the zone. The military made us watch while they cut up our IDs. When they let us go, we marched to the Church, where we had to try and plan what to do next. We went from house to house collecting all the Lotus workers. At midnight we moved to get back into the zone. We did it crawling through the drains under the fence and into the zone. Yes most of the women did this. I was frightened, but we were all together. Inside the zone we hid ourselves in the darkness and waited. Then at about 7.00 a.m. when the factories opened for the morning shift, when workers came into the zone to start work, we came out of our hiding places and set up the picket line again. . . . Back up went our banners. The military were amazed and angry.[47]

Despite severe coercion the workers won their demands.

Consciousness developed at a tremendous speed in the 1980s when a community spirit at the factory level was created by three general strikes which paralysed operations in the Bataan FTZ. The first of these 'sympathy strikes' was in June 1982, when 15,000 workers from twenty-three factories walked out to support the workers picketing against police brutality and the arrest of workers at Inter Asia Container Inc.

In October 1983 the second general strike took place, in which around 20,000 workers participated. Here again workers supported the picket line at the British Astec electronics company, protesting against military attacks. In May 1984 another violent encounter with the state occurred. According to the bulletin *International Labour Reports*,

> workers picketing at Interasia Co., a Japanese owned company in the zone, were attacked on the picket line . . . workers were beaten with truncheons and kicked before being hauled inside a bus hired by the Management. While on the way to the Police headquarters, workers managed to stop the bus by grabbing the wheel from the driver. On the following day a *third* general strike was launched, bringing 90% of production in the zone to a halt.[48]

Strikes and protests were still taking place in the Garment Manufacturing Corporation in the Bataan FTZ in 1989–90. Throughout the decade the workers' consciousness was being influenced not only by grievances in the FTZ but by wider political circumstances. Women workers have been part of this process and their unfulfilled economic and social expectations were to find a militant expression. The support given by a labour movement which displayed commitment and the ability to withstand heavy repression contributed to a feeling of being together as workers and women. This generates new beliefs and the strength to resist. The great contradiction which this militant consciousness had to face by the end of the tumultuous decade was that its very success could result in the withdrawal of companies from the FTZ. For, despite the collective power they were able to assert against the odds, Filipino workers have been nonetheless dependent on international capital which has been inclined to search for less militant sites of labour.

COMMUNITY ORGANISING

Along with women's organisation through trade unions, women's groups also made contact with women workers through community-based organisations during the 1980s. For example, in Sri Lanka various organisations set up centres directed at women workers. They have differing reasons for working in these areas and therefore have a range of approaches and ideologies. Nevertheless, they have faced similar problems, including repression from the state, and have found ways to work together on specific issues.

Christian organisations, both Catholic and Protestant, have formed a distinct and important part in the response to the needs of workers in the FTZ. Both Catholic and Protestant groups have worked with women workers since the inception of the FTZ in 1979 by separately initiating various facilities for women. These have fostered activities which attempt to integrate women into the village communities, creating a channel by which they have been able to educate the women about their rights.

There are seven Catholic centres in the villages adjoining the FTZ. Some of them provide boarding house facilities for the workers, and in some nuns are available to lend a sympathetic ear to the problems of the women. Space is provided where the women can meet to discuss and socialise. Library facilities are available in some of the centres while seminars or social activities are often organised.

A food co-op was initiated by one of the Catholic centres, to combat the rising prices of food, for which the villagers blamed the women. The women workers bought shares to build up sufficient capital to set up the co-op. The result was the lowering of prices of basic food items, due to a high degree of patronage by the villagers and FTZ women which forced the other shopkeepers to bring their prices down. The experience demonstrated to the women workers that it was possible to develop strategies together which proved beneficial to them and the villagers, while also winning the confidence of the community.

There has also been collaboration between the Christian churches and the Buddhists. In the early 1980s an alliance of women's groups, the Legal Advice Centres and the Buddhist organisations campaigning to stop the sexual harassment of women workers returning after work brought the active participation of all to the fore. The women made posters, pamphlets, and did door-to-door campaigning themselves. The message was preached in every church and Buddhist temple. Petitions were sent to the Greater Colombo Economic Commission, requesting greater security and the provision of transport for the women. These experiences caused the women and the active organisations to come closer together. The campaign was able to bring sensitive and hidden subjects into the open and also legitimated the relation of outsiders to the workers' problems.

Women workers themselves do not control the running of the Catholic or the Protestant centres. However, they do participate in the committees where decisions are taken, though key persons who are not themselves women workers sit in to guide the discussions. Allowing for certain limitations because of religious ideology or their answerability to 'higher' bodies, religious centres nonetheless have been able to provide women with places which they can rely upon and to which they can relate.

Legal advice is provided by nearly all these centres. However, a specialist Legal Advice Centre gives free advice as well as intervening on behalf of workers – either free of charge or at a minimum basic charge. The lawyers and full-time personnel write and distribute pamphlets on legal issues to the workers, while seminars and discussions are organised where workers are invited to participate. It plays an important role in a situation where the basic rights of workers are vulnerable. From the late 1980s the Legal Advice Centre came under very heavy repression. Full-time personnel suffered physical attacks, and some have even disappeared.

A Women's Centre was set up by a group of women who were formerly factory workers. It has offered FTZ workers a multifaceted programme with diverse facilities which include legal and medical assistance, library facilities and training in alternative skills. It has conducted study seminars and discussions, providing education for women on their rights. Due to its focus on women, and its convenient location with close access to the FTZ, the Women's Centre has been extremely popular.

Autonomous community initiatives have also occurred in Malaysia. Two examples are the Sahabat Wanita Friends of Women (FOW) and Tenaganita Women's Workforce Action Group (WWAG). Both have combined educational and social services providing practical assistance as well as building confidence. FOW has dedicated itself to promoting the rights of women working in factories as well as on the plantation estates. They have run kindergartens in areas where the women live, which have resulted in widening contacts. The group has run courses and seminars for women from large and small factories, in an attempt to raise their consciousness about issues such as health and safety. The Tenaganita Women's Workforce Action Group was established in the late 1980s and is located in a house bordering Penang Bayan Lepas Free Trade Zone. Tenaganita is run by a worker leader from the struggle against the retrenchment of women at Mostak. She attempts to relate her religious belief, that of a 'liberated Islam', with the needs and welfare of workers.[49] She affirms that Tenaganita is open 'to all female (and male) factory workers regardless of race, and Islam is not incompatible with trade unions or workers' organisations'.[50] Elizabeth Grace reported in 1989 that Tenaganita intended to offer various courses in English, typing, sewing and public speaking for women workers, as well as providing short-term accommodation to women who have migrated to Penang looking for work.

All these organisations have tended to be *for* women workers rather than *of* women workers as they have been mostly set up by external groups. Nevertheless, the centres have run considerable risks to work in the areas of the FTZs – sensitive areas where police victimisation is rampant. Their members have not simply provided useful services but faced real dangers in order to seek justice for the women workers employed in the zones.

Since 1984 in Sri Lanka the publication and distribution of a newspaper has been another kind of intervention made by an autonomous women's group. This has provided a medium by which the FTZ women workers have learned to express themselves and to share their ideas with other women. It also has created a vehicle through which they were able to state their views to employers or those in authority while mobilising support for their demands. The collective which produced the FTZ newspaper has been active for many years in various organisations around the FTZ. The newspaper has included articles, poems and stories sent in by the women workers themselves, while the collective has written linking articles, which have given advice, news of campaigns or the activities of other organisations in the zone. The newspaper has had a circulation fluctuating around 8,000 since its inception. During the initial five years the paper was distributed free of charge. This enabled it to become known and popular. From 1989 it was sold across the boarding houses, or along the streets where women go to work. On many occasions the women's newspaper made its way into the factory, although workers were officially not allowed to take it inside. Inside the factory it travelled down the assembly line or someone might pin it up on the noticeboard.

The response to the newspaper from the managements within the FTZ has been diverse. While some take great pains to find out who gave out information related to the factory (and sometimes reprimand them), others find it a useful method by which to obtain information about themselves and the views of the workers and their expectations or grievances. Sometimes managements have sent responses to articles appearing in the paper, which is published in the next issue giving all the workers a chance to see this response.

The FTZ newspaper has been an important campaigner on issues such as night work, sexual harassment and problems relating to organising. It has also taken up the difficulties of women travelling to and from work.

CONCLUSIONS

In all three countries direct as well as covert methods of organising have been undertaken. There has been a marked tendency to approach the women not simply as workers employed in factories but to take into consideration their lives in the boarding houses, on the streets and in the community as well. New forms of organising will undoubtedly continue to appear through the interaction of the clandestine and open efforts of women workers to resist.

A newly emerging consciousness can be seen among women workers in what is a relatively recent phase of industrialisation. The employment of the women in the FTZs has meant that very basic changes have occurred in the lives and attitudes of these new workforces. For example, many women have come to realise that their employment in FTZ factories is not to be as brief as they originally expected but will occupy most of their youth. Therefore they can be seen consciously or unconsciously resolving to make it as liveable as possible for themselves and others. It has also become apparent that marriage is not necessarily a way out of their jobs, for many women who married have still had to continue to work in the FTZs. Consequently, they have been forced to recognise that returning to their villages and resuming their former lives before employment in the zone is not as easy as they once thought. The FTZ experience has left its mark on cultural patterns as well as upon the women as individuals.

Facing the reality that they may be in FTZ factory employment for quite a while, and confronted by the difficult conditions of work and living, has not, however, completely destroyed the dream of a better life for themselves and for their families or a fairer deal on the job. The conflicts and contradictions between dreams and reality in their daily lives has combined with the collective understanding which has come from organising their lives together with other women. From Sri Lanka Rohini described in 1990 how she gained the power of self-assertion and a sense of solidarity with other women.

Anyway working in the FTZ was an experience worth to have had. I grew up a lot during that period. I myself can't see any changes in my appearance, but I think you can see that better! But I grew in my ideas, what it is to fight for what I believe. At home I was protected by my parents. We had no idea of being responsible towards each other. But now it is different. Now I know what suppression is. What repression is. I also know that I have to stand against it.

I like the freedom I got through this job. The freedom to make my own decisions, to spend money, to go to places I like to be. It is also a hard life. There were many days I felt I could not cope. Money was often scarce. I was often hungry . . . but we learnt to deal with such situations. I sometimes imagine to go back to the village. I know that it won't be easy. Now that the FTZs have been established in Sri Lanka I could say that there are good and bad sides to it. From my personal experience I know that I grew and learnt about being a woman worker and to live together and be responsible for other women. But the state and authorities should take more responsibility for the thousands of women who leave their villages and come to work in the FTZ. They should ensure that decent wages are paid. That transport, and security, is provided, that decent living conditions are ensured. That the right of women to organise is protected. They should take our side. FTZs have provided us jobs. That is true. But we also want to feel secure and justly treated.[51]

Her statement indicates the extraordinary transformation which capital has effected in translating the life of a village girl into a FTZ worker.

The diverse forms of organising are mutable and dynamic. They have emerged out of a new form of industrialisation and they will continue to adapt and reconstitute themselves as circumstances and consciousness shift. Rather than simply studying the movement of capital, the emergence of a new awareness both as women and as workers needs to be much more closely examined in specific FTZs.

By looking at the consequences of employment in the FTZs from the perspective of the women whose lives have been so deeply affected, it is possible to subjectify the argument put forward by Diane Elson and Ruth Pearson as early as 1980, and avoid a deterministic approach. The recent history of FTZs shows that, indeed, 'the relation between capital and labour can never be completely determined by what would suit capital.'[52]

In the words of Manike, a woman worker who was dismissed from the Sri Lankan FTZ for writing a poem in the workers' newspaper,

I awake early morning at 4.30 a.m.
I have to kindle the fire
Having washed my face, I gulp down some tea
I leave for work early morning.

I start work at 7 a.m.
The supervisor demands the production

I regret my inability to meet this target
She scolds us for this.

At 10.30 a.m. we get a sip of tea,
The tea contains no flavour, no sweetness
We drink it to quench our hunger
We tolerate these because we are poor.

I came to Katunayake because I was without work
I came to the Free Trade Zone to work
I worked at Star Garments
Now I am tired and disgusted with the job.

The other day I fell sick,
But I was not allowed to leave the factory
I know that one day I will have to work
– even through sickness,
I will surely fall dead, at Star Garments.

I work throughout the month
I am paid Rs 800/- for the month
An attendance bonus of Rs 72/- is paid
We are paid with no further allowances.

At 7 a.m. I sit at the machine
By 8 a.m. the supervisor is already at my side
She asks me what my production is
I tell her only the amount I can give.

I often get a pain in my chest
The supervisor asks me to go to the sick room
I can stay there around quarter of an hour
I come back again and sit at the machine.

My mother does not know how much I suffer,
Only I know how much I suffer,
I leave in the morning and come back at night,
I suffer with the pain in my body.

We are not given any leave,
Leave is allowed only in emergencies,
That leave is also granted after much argument,
We who are poor are made to suffer so much.

My mother who fed me with her own milk,
My father who worked so hard to bring us up,
My teacher who gave me the knowledge,
To them I pay my respects.[53]

APPENDIX 1

Incentives offered by the three countries, namely Malaysia, the Philippines and Sri Lanka, to foreign and local investors seeking to invest in the FTZs:

Malaysia

Full or partial exemption from relevant laws and regulations;
tax exemptions (up to sixteen years);
no restrictions on repatriation of profits or on imports;
low rentals;
cheap labour;
laws restricting strikes and ensuring production.

The Philippines

100 per cent ownership;
no duties, taxes or licence fees on imports to the zone;
the right to borrow within the Philippines with government guarantees for foreign loans;
no taxes on exports;
no minimum investment requirement;
unrestricted repatriation of capital and profits;
freedom to sell 30 per cent of the annual output of the FTZ firms on the local market (a means to avoid quota on imports).

Sri Lanka

No limit on the equity holdings of foreign investors;
free tranfer of shares within and outside Sri Lanka;
no tax or exchange controls on such transfers;
dividends to non-resident shareholders exempt from tax and from exchange controls;
unrestricted remittance of capital and proceeds of liquidation;
100 per cent tax holiday for up to ten years, depending on the number of employees, the amount invested (fixed capital) 'and the levels of foreign exchange earned on export sales'.

(Information from APHD, 1986)

NOTES

1 Quoted in 'We Can Fight', International Labour Reports, Manchester, July/August 1985.
2 See, for example, Susan Joekes, *Women in the World Economy: An INSTRAW Study*, Oxford University Press, New York, 1987; Gillian H.C. Foo and Linda Y.C. Lim,

'Poverty, Ideology and Women Export Factory Workers in South-East Asia', in Haleh Afshar and Bina Agarwal (eds) *Women, Poverty and Ideology in Asia*, Macmillan Press, London, 1989; and Noeleen Heyzer, 'Daughters in Industry', Asia Pacific Development Centre, Kuala Lumpur, Malaysia, 1988.

3 See Diane Elson and Ruth Pearson, 'The Latest Phase of the Internationalisation of Capital and its Implications for Women in the Third World', Institute of Development Studies, Sussex, Discussion Paper 150, June 1980, p. 9.
4 See Asia Partnership for Human Development (APHD), 'Export Oriented Industrialisation in Five Countries – Philippines', in Dennis Shoesmith (ed.) *Export Processing Zones in Five Countries – The Economic and Human Consequences*, Asia Partnership for Human Development, Hong Kong, 1986, p. 44.
5 ibid.
6 International Labour Organisation and United Nations Centre on Transnational Corporations (ILO/UNCTC), *Economic and Social Effects of Multinational Enterprises in Export Processing Zones*, ILO, Geneva, 1988, back cover page.
7 See Asia Partnership for Human Development (APHD), 'Export Oriented Industrialisation in Five Countries – Malaysia' in Shoesmith, *Export Processing Zones in Five Countries – The Economic and Human Consequences*, pp. 53–4.
8 APHD, 'Export Oriented Industrialisation in Five Countries – The Philippines', p. 44.
9 See *Central Bank Annual Report*, Central Bank, Colombo, 1990, p. 57.
10 Voice of Women (Kantha Itanda), *Women Workers in the Free Trade Zone of Sri Lanka*, A Survey of Sri Lanka, Colombo, 1983.
11 See Kumudhini Rosa, 'Women Workers' Strategies of Organising and Resistance in the Sri Lankan Free Trade Zone (FTZ)', Institute of Development Studies, Discussion Paper 266, Sussex, 1989, p. 8.
12 APHD, 'Workers and Their Communities: The Human Consequences', in Shoesmith, *Export Processing Zones in Five Countries – The Economic and Human Consequences*, p. 91.
13 Quoted in Rosa, 'Women Workers' Strategies of Organising and Resistance in the Sri Lankan Free Trade Zone', p. 3.
14 APHD, 'Workers and Their Communities: The Human Consequences', p. 98.
15 APHD, 'Workers' Response and Workers' Organisation', in Shoesmith, *Export Processing Zones in Five Countries – The Economic and Human Consequences*, p. 116.
16 ibid.
17 ibid., p. 177.
18 Quoted in Rosa, 'Women Workers' Strategies of Organising and Resistance in the Sri Lankan Free Trade Zone (FTZ)', p. 4.
19 APHD, 'Workers' Response and Workers' Organisation', p. 113.
20 See Committee for Asian Women (CAW), 'Philippine Women Workers Fought Anti-Union Labour Law', *Asian Women Workers Newsletter*, 8(4), Hong Kong, December 1989, p. 10.
21 Rosa, 'Women Workers' Strategies of Organising and Resistance in the Sri Lankan Free Trade Zone (FTZ)', p. 5.
22 Leela quoted in ibid., p. 3.
23 ibid., p. 4.
24 Greater Colombo Economic Commission, *Brochure for Investors*, Greater Colombo Economic Commission, Colombo, n.d., p. 8.
25 Rohini quoted in 'Listen to Me', *Drops of Sweat*, 3, 1990, Pinneberg, Germany, pp. 4–5.
26 APHD, 'Workers and Their Communities: The Human Consequences', pp. 90–1.
27 Elizabeth Grace, *Shortcircuiting Labour – Unionising Electronics Workers in Malaysia*, INSAN, Kuala Lumpur, Malaysia, 1990, p. 46.

28 ibid., p. 46.
29 ibid., p. 46.
30 ibid., p. 47.
31 See James Lochead, 'Retrenchment in a Malaysian Free Trade Zone', in Noeleen Heyzer (ed.) *Daughters in Industry*, Asian Pacific Development Centre, Kuala Lumpur, Malaysia, 1988, pp. 282–3.
32 APHD, 'Malaysian Case Study', in Shoesmith, *Export Processing Zones in Five Countries – The Economic and Human Consequences*, p. 178.
33 ibid., p. 113.
34 ibid., p. 115.
35 ibid., p. 217.
36 Grace, *Shortcircuiting Labour, Unionising Electronics Workers in Malaysia*, p. 2.
37 Malaysian Trade Union Congress (MTUC), Letter to the ILO: 'MTUC Petitions the ILO! No Union for Electronic Workers', *Aliran Monthly*, 9(3), Penang, Malaysia, 1989, p. 20.
38 ibid., p. 22.
39 Grace, *Shortcircuiting Labour, Unionising Electronics Workers in Malaysia*, p. 43.
40 ibid., p. 56.
41 ibid., p. 47.
42 ibid., p. 47.
43 ibid., p. 47.
44 Kilusan ng Manggagawang Kababaihan (KMK), *Beyond Labour Issues – Women Workers in Asia, Case Study: Philippines*, Committee for Asian Women, Hong Kong, 1988, p. 46.
45 ibid., p. 16.
46 ibid., p. 46.
47 Lisa, quoted in 'We Can Fight', p. 16.
48 'The New Challenge', *International Labour Reports*, September/October 1984, Manchester, p. 18.
49 Elizabeth Grace, 'Unionisation in Malaysia's Electronics Industry: In House Unions and Women Workers', Institute of Development Studies, Sussex, M.Phil. Dissertation, 1989, p. 50.
50 ibid., p. 51.
51 Rohini quoted in 'Listen to Me', *Drops of Sweat*, pp. 4–5.
52 Elson and Pearson, 'The Latest Phase of the Internationalisation of Capital and its Implications for Women in the Third World', p. 9.
53 Manike, 'Life', quoted in Rosa, 'Women Workers' Strategies of Organising and Resistance in the Sri Lankan Free Trade Zone (FTZ)', pp. 23–4.

Chapter 4

Weaving dreams, constructing realities

The Nineteenth of September National Union of Garment Workers in Mexico

Silvia Tirado

ABSTRACT

Alongside the rise of highly technologised flexible units which produce for specialised markets, some industries, particularly clothing in the Third World and in pockets of the First, have continued to use labour-intensive methods. These have had an existence which is often clandestine. During the 1980s these casualised forms of production not only proliferated but were legitimated by economic policies which removed restrictions upon them.

The Nineteenth of September Union arose after the 1985 earthquake had devastated Mexico City. Garment workers in the sweatshops were buried alive under the rubble. Their bodies were found amidst the wreckage.

Silvia Tirado interviewed women workers in sweatshops before the earthquake and has worked closely with the Nineteenth of September Union which was able to unionise unorganised women workers and also gain official recognition for a union independent of state structures.

While tremendous economic and social obstacles have restricted the union's growth and capacity to act, it remains a significant model, being not simply autonomous but raising demands which relate to women's social needs.

The second half of the 1980s saw the collapse of social policies based on need internationally. This was caused not so much by their theoretical errors as by their mistaken and ineffective implementation. The prevailing belief in a market-oriented approach inspired a global capitalist restructuring with the promise that it would bring renewed productivity, growth and profitability. This restructuring has been marked by regional integration in the international sphere. The emergence of three great trading blocks – Europe, Japan and Asia, and the United States with Canada and Latin America – has forced each country to combine with the others, and to adopt a code of behaviour in accordance with its status within this international division of labour.

The Latin America countries formed a continental trading block. Each country has adopted economic codes imposed by the great financial institutions which include the commitment to modernise by privatising strategic or profitable sectors formerly controlled by the state. Mexico has distinguished itself for its 'good conduct' within these institutions, yet the cost of fulfilling such programmes has fallen upon the living conditions of workers and their buying power has declined to unprecedented levels. Factories have closed, the achievements won by workers through years of struggle have been disregarded and independent and democratic movements are being repressed. As a result many workers are receiving pay which barely allows for subsistence, while others have become unemployed.

This process of modernisation has involved various industries in different ways. New technologies have been introduced into certain strategic industries and have significantly displaced labour. In contrast, the garment industry has been left behind by recent technological changes. The sector continues to be characterised by: the intensive use of labour; limited capital intensity in machinery; little mechanisation or automation; factory organisation which is increasingly Taylorist-Fordist.

New technology has developed in parts of the productive process: in designing and cutting (grading, sketching, cutting of moulds and pieces of cloth) and in areas of accounting, administration and inventories. Despite their restricted use thus far, the impact of new technologies remains important in the garment industry internationally, especially in countries like North America with higher wage levels.

In Mexico, however, robotics and micro-electronics have been introduced only in partial processes in the large garment firms. These technological innovations are generally limited and diffusing slowly. This is partly because the garment industry has several peculiarities. The type of clothing which is in demand varies according to season, temperature, style and the economic level of the consumer. So while products most prone to standardisation, such as shirts and trousers, are manufactured in large firms which make greater use of new technologies, many fashionable goods have fluctuating markets which makes employers opt for the flexibility of units which require little capital.

Another characteristic of the industry is its intense utilisation of labour which is why it often locates itself in areas and regions where abundant and cheap labour with low skills is available. The businessmen in this industry display a 'conservative' attitude towards transforming the productive plant, preferring to continue producing in conditions which are comparable to those found at the beginning of the century. Productivity is low and efforts to increase it are likely to seek to intensify the pressure upon labour rather than to introduce new technology.

Owing to the relative facility with which companies can enter the industry which does not always require large sums of investment capital, a vast number of small workshops, often family-owned and operated, and a large number of clandestine home-based workshops, assemble parts of garments. These clandestine operations can only survive by evading the minimum legal rights of workers. Owners avoid the additional expenses involved in installing a factory, evade taxes and often even get out of purchasing the necessary machinery, since most tools belong to the garment workers themselves.

Ninety per cent of these garment workers are women. In addition to this being considered 'women's work', women are employed because of their lack of professional training which means they can be paid low wages. Working conditions in the garment factories are poor; instead of the eight hours set by the law the actual work day is between ten and twelve hours, and some women work on Saturdays. Wages are based upon a piece-rate system, which intensifies the pace of work as each minute signifies the gain or loss of additional pesos; exhaustion only results in losses as the worker is forced to slow down.

The majority of workplaces in the clothing industry are deathtraps. In addition to always being crowded with stock and utilising highly flammable material, they lack safety measures. There are frequently no fire-extinguisher systems, or first aid kits in case of accidents, or emergency exits in case of disasters. In addition to the lack of adequate safety provision in these 'mousetraps' (as workers call these workshops), management also engages in the reprehensible practice of locking women workers inside the factory while they work, leaving them defenceless against any emergency which may arise.

When workers are unionised, it is usually within official company unions, making it difficult for workers to change working conditions which are neither favourable nor just. In order to divide and put pressure on workers, management utilises diverse strategies, often violating the legal rights which have been granted to workers. In the majority of firms, workers are poorly treated and report experiencing sexual harassment, physical aggression and favouritism by management and administrators. In the words of one garment worker:

> 'Violations against us as people are innumerable in the garment workshops. If these deeds were punished, no garment workshop owner would be spared. We would have to denounce hundreds of offences, unjustified punishments and insults and mental torture. Examples range from losing one hour of pay for arriving one minute late, to more harsh actions, such as making a young woman walk in a circle 80 centimetres in diameter during the entire working day'.

Another woman described cruelty to a worker the employers no longer wanted:

> 'They made Esperanza weave pure black cloth for one month, just to bore her, to tire her. Then they took away her sewing machine and made her sit the entire day in a chair. Then they took away the chair and she stood for two weeks. Finally, they told her that there was no work available. They placed a spool of thread on top of a pole, which she had to unwind and re-wind and counting for hours as she did so, she became exhausted and left.'

These personal examples of harsh and unjust treatment have to be understood within the specific circumstances of Mexico's labour relations. For more than a century, the Mexican working class has played an important role in the nation's history; no important political-economic event has been free from the influence of workers' aspirations and struggles. Official ideology continued to be phrased in defence of workers' rights; in practice, however, this discourse has been manipulated, leading to divisions in the labour movement which only benefit management and the state.

Mexican unions are either affiliated to government organisations, or they are independent. Official union leaders are firmly linked to the state and sell workers' collective bargaining agreements to management, permit unjustified dismissals, and acquiesce with the interests of management. Independent unions always confront great difficulties, ranging from achieving legal recognition of their organisation, to acceptance of their claims. Hence, the great worker and peasant unions are essentially state corporations, led by corrupt leaders, who subordinate the interests of their members to a cynical rhetoric which no one believes in any longer.

Mexican labour unions are faced with a political economy of 'modernisation' which consists of relinquishing the reins of the economy to the free 'competitive' market, yet pits the country against the much more developed economy of the United States. Currently, official unionism is divided over whether to struggle for increased salaries or to defend the legislative achievements obtained by the working class in the past which are now under threat.

In order to increase the competitiveness of the productive capacity of firms, making them more attractive to foreign capital, both management and the state want to eliminate the framework of legal work regulations and allow management and workers to determine work conditions in a direct manner, according to 'the current viability of the firm'.

Under the banner of modernisation, they aim to pay by the hour, abolishing any guarantee of a fixed work day and salary. The payment of pensions and benefits would be left up to each firm. To obtain a flexible workforce, they have used seasonal contracting, instituted greater salary differentiation and increased the power to dismiss workers on annual contracts, in order to create a so-called 'multiskilled' worker who can undertake multiple activities as cheap and flexible labour. However, it is only possible to achieve these objectives if the unions have been weakened and their scope for action has been limited.

These developments raise serious questions for the future perspectives of waged female labour, which has grown significantly since the early 1970s. Will female workers be subjected to the same process of modernisation as male workers? Do labour programmes include specific demands regarding women workers? What has been the role of women in the unions?

Women workers have a specific position distinct from that of male workers within production because of the historical subordination of women not only in the workplace, but in society and the home. Occupational segregation continues to be profound and persistent. But the economically active female population has grown more rapidly than among male workers. In absolute terms, the number of working women in Mexico grew from 2.035 million in 1960, to 2.466 million in 1970, and reached approximately 6 million in 1980.

The sectoral distribution of female labour also presents a growing tendency to move into the informal sector. In 1970, 33 per cent of women in the economically active population were working in the primary sector, only 14 per cent of women workers were in the secondary sector, and 53 per cent worked in the tertiary sector. By 1979, these same figures had altered to become 22 per cent, 4 per cent and 72 per cent respectively.

The contemporary process of industrial restructuring within the Mexican economy affects women and men differently, yet many of its future consequences remain unclear. The traditional role of mother/wife/housewife has been redefined for a large proportion of adult women, as they have found it necessary to raise income for family consumption through different activities. Paradoxically, as certain sectors of employment are being closed to men, occupations have opened for women. One reason is that female labour is cheap and unskilled. Changes in the structure of production and the rise in the cost of living which puts pressure on consumption have contributed.

Opportunities for work in the formal sector have moreover diminished considerably and the informal economy is increasing. A projection by the government, covering 1986 to 1990, calculated that in Mexico City alone, 7.8 million people worked in the informal sector, out of a total population in this urban metropolis of 28 million. These statistics also indicated that the participation of women in the labour market will increase. In 1989, the economically active female population was 33 per cent, showing an increase of 5 per cent from 1980. The cost of the basket of basic goods had also increased considerably, including sugar, bread and milk, as well as other items not under control, such as meat, fruit and vegetables. The policy of wage restraint and constant price increases have made women and children take on paid labour in order to help the family survive.

Another factor encouraging the incorporation of women in formal and informal labour is the increase of women-headed households. According to data from the family courts, 50,000 divorces were registered in 1989 for every 100,000 marriages – in other words, half of all marriages end in divorce. This phenomenon is particularly high among women garment workers, which means that it is these women who are ultimately responsible for their children and for the care and survival of the family as a whole.

These single, separated, divorced or abandoned women can only count upon their labour to support their children and other members of the family, such as parents and younger siblings. Under this crisis situation, a woman's salary is vital for the subsistence of the family, and not merely as a 'complement' to the male income.

As existing inequalities and the process of economic reorganisation have intensified, it is now considered 'a blessing' to have and keep a job. The challenge which currently faces workers in general, and women workers more specifically, is how to respond in this transition period to economic restructuring and increased productivity, without undermining what has already been gained.

The wider social circumstances of Mexican society have made this dilemma increasingly urgent. The incorporation of women in the production process has not liberated them from the double work day, since women not only contribute a salary for the family's survival, but are also responsible for domestic work, the family's shopping and other household tasks. The state does not subsidise any public services. The lack of laundries, canteens, creches and pre-prepared foods makes a woman's domestic tasks much greater. Women workers need to work long hours during the day. Exhaustion and nervous illnesses not only reduce their productivity at work but also bring about an irreversible process of personal deterioration. Childcare is a pressing problem, yet the state budget for creches is very limited. In 1990 only 8.5 per cent of women workers left their children in creches. Therefore, more than 50 per cent of working mothers had to find an alternative solution for the care of their children during the work day. As a result, many women leave their children in the house of a family member, or with neighbours, or in extreme cases mothers leave their children locked up in the house during the work day.

Motherhood is subject to labour discrimination. Aside from being the principal reason for not being hired, in other cases the minimum rights which the Mexican laws provide for mothers and pregnant women are violated. In the industrial sector, and particularly in the garment industry, it is common for a woman to undergo a medical examination to determine if she is pregnant, despite the fact that this practice is prohibited by the Federal Labour Laws.

Violence against women in the home and on the street has been getting worse; the number of sexual violations is scandalous, although only a small proportion of such incidents are actually reported. Sexual harassment in the workplace is also a daily problem, yet few women take recourse to the law, for they run the risk of losing their job.

The official unions have hardly begun to confront the changing situation and the specific problems it is posing for women. The Mexican unions have never included any strategy or even specific policies for women in their labour regulation programmes. The rights of women workers in the Federal Labour Code are specified in Articles 164, 169, 170 and 171 – yet, it is worth noting that they refer fundamentally to women workers in their role as mothers and the rights they derive from this role. For example, Article 164 refers to inequalities regarding the rights and obligations of men and women workers, yet in practice this article is ignored. Other important issues such as sexual harassment, sexual segregation in the workplace and disqualification of female labour have not been addressed within the existing labour codes.

Women workers in the unions are caught within a power structure where the majority of leaders are men, and these men determine the objectives and procedures for action. This pattern is observed even in unions where women comprise half of the membership (in occupations such as telephone operators, bank workers and university staff).

The exception is the Nineteenth of September Garment Workers' Union which is composed solely of women. Established in 1985, the Nineteenth of September Union recruited from the sweatshops of Mexico City. The example of garment workers unionising is one of the most significant within the independent labour movement because it represents an alternative approach to organisation which links workplace and social demands, and because it is one of the few independent unions which have been officially recognised. It is autonomous and democratically run.

THE GARMENT WORKERS ORGANISE THEMSELVES

The movement among women garment workers was an unexpected development in Mexican labour organisation. The women worked in small and unhealthy workplaces, remained seated for more than twelve hours per day, were subjected to the pressure of piecework and the infernal noise of the sewing machines, yet earned miserable wages. After decades of work under intense exploitation in this sector, it took the tragedy of the earthquake in Mexico City in 1985 to open the eyes of society to the conditions in which these women lived and worked. It is a disgrace that it required scenes of bodies crushed under the debris and between rolls of cloth and machinery, or of insensitive owners trying to recover their equipment in the midst of the tears of the garment workers' families, to raise awareness about such an important, yet so marginalised, sector of the economy.

The survivors, who had witnessed the walls of their factories fall upon their places of work, were sacked by their bosses without any explanation and without regard for the existing labour laws, principally the law of seniority. The garment workers were indignant at the indifference and negligence displayed by both businessmen and politicians to the situation of hundreds of women workers who had been crushed and killed under the concrete walls. Employers removed machines leaving the bodies of the dead in the rubble. More than 5,000 garment workers united and organised themselves – first to demand the rescue of their fellow workers, and afterwards to defend their rights against the injustice of bosses who had paid no compensation to any workers, living or dead. The garment workers, who had always worked in isolated workshops, recognised in the trauma of the earthquake that each one of them had suffered the same forms of oppression and decided to organise themselves independently.

They had been affiliated to company unions, which had colluded with the owners and only served to institutionalise their exploitation. This disaster drew public attention to the harsh system of production to which garment workers had been subjected – working long hours for less than the minimum wage set under the Mexican Federal Labour Laws. Businessmen were not only guilty of conscious neglect of labour laws and appalling working conditions; their practice of keeping workers locked in the factories had prevented many from escaping when the earthquake struck.

The catastrophe also revealed that most garment workers did not even know to which union they belonged, making it impossible to solicit union intervention in the solution of conflicts. Given the character of the official unions and the ignorance of most workers about their official union, many workers sought assistance from a group of democratic and feminist lawyers who had provided legal assistance to them previously. The moment proved decisive, for under any other circumstances, their struggle could have been absorbed within one of the official company unions. It would have been inconceivable before the earthquake that the solidarity of the garment workers could have merged and grown, providing the strength to construct a combative organisation able to confront businessmen and government authorities.

On 20 October 1985, the year of the earthquake, these workers succeeded in being recognised as the National Union of Garment Workers, or the Nineteenth of September Union. Their first congress was held in May 1986.

The formation and legalisation of the garment workers' union in 1985 was a great achievement for, since 1976, no democratic union had been legalised in opposition to the official unions. Challenge to the power and control wielded by the government and by the Mexican Labour Confederation – the largest official labour organisation in the country and the one to which most workers are linked – was possible thanks to solidarity nationally and internationally and to a rising social movement in Mexico which had broken fundamentally with the existing system.

The most difficult test for any independent and democratic union is to confront the officially incorporated structures which have protected the owners for decades. The Nineteenth of September Union was able to make the conditions in the garment industry public while achieving autonomy from the company unions, the garment businessmen and the government authorities. A key factor for an independent union is the right to negotiate the women's collective work contracts. The fifth congress in September 1990 reconstituted Article IV of its statutes and with this, transformed the conditions under which workers affiliate themselves to the Nineteenth of September. As a result its members gained the same rights as any other worker whose collective contract is held by a union. This meant that the Nineteenth of September Union did not have to expend energy fighting to gain the right to negotiate work contracts.

The union is unique because it is a democratic organisation run by women workers. The congress is the highest authority of the union and elects the National Executive Committee as well as having the power to reform the statutes if one does not function well. There are eight Secretaries on this Executive apart from the General Secretary. They are responsible for Labour, Disputes, Personnel Relations, Finance, Press and Propaganda, Political and Union Education, Acts and Agreements, and Sports and Leisure. Each woman holds her post for two years.

In the factory or workshop, two other organisational forms contribute to the life of the union: the executive shop steward council and the shop stewards assembly. The shop stewards are elected by women workers in their general assemblies which are held each month. They all participate in the decisions and are aware of issues under negotiation. The leadership of the union is entirely composed of women. These representatives are elected by secret ballot by the entire membership. Each group of ten garment workers comprises one section which elects its own delegate. The union fees are 2 per cent of the women's salary.

The union's demands relate both to the particular conditions of the clothing industry and to the critical economic situation which exists in the country. In the first Special National Congress held on 17–18 May 1986, the Nineteenth of September Union approved the following demands:

1 A call to all the women in the sector to struggle for the respect of the professional minimum salary as set by the Federal Labour Law.
2 That all women workers be informed of all the collective work contracts, thereby enabling their direct participation in contractual revisions.
3 That the authorities no longer permit the existence of clandestine workshops, that their existence be regulated legally and that the labour rights of workers in these workshops be respected.
4 A policy which prevents women workers from being exploited as workers.
5 That the work centres in this sector provide adequate hygiene and security facilities for workers.
6 That the authorities make public the existing contracts which protect the shop owners and annul them immediately.
7 That the true nature of the co-operatives promoted by the government should be exposed and labour–management relations established with consequent unionisation and a collective work contract.
8 Setting up of creches, the right to free maternity leave, freedom from sexual harassment and violence against women, respect for motherhood in the workplace, for pregnancy to be eliminated as grounds for dismissal or refusing employment to a woman, and for full equality between women and men.
9 The suspension of payment of the external debt.

Since its formation in 1985, the National Garment Workers' Union, the Nineteenth of September Union, has presented an important model for garment workers, showing it is possible to struggle for just and equitable working conditions and obtain dignified treatment for women workers. In the first four months of its existence, the Nineteenth of September Union succeeded in forcing eighty businessmen to compensate more than 8,000 women workers who had lost their employment due to the earthquake. This achievement encouraged many other garment workers to affiliate to the union and to work within the organisation. The union also began to organise garment workers actively in factories to secure their own legal representation directly. In its first months of existence, the union grew to include more than 3,000 affiliated members.

In the Red Heart factory, where the union won its first collective agreement on 17 December 1985, the factory owner was in alliance with the Mexican Labour Confederation, and dismissed twenty-three women workers who had proclaimed their support for the new union. Thanks to the consistent and combative nature of these women workers, they continued to win new work contracts under the new workers' union.

Nevertheless, the women garment workers have suffered significant defeats and are confronting formidable obstacles. On one occasion workers arrived on the day of elections for their union representative, only to find that the factory had been occupied by armed men from the Mexican Labour Confederation. This occupying force introduced fifty false votes from people who did not work in the factory. Although the elections were declared fraudulent, the authorities claimed that the Mexican Labour Confederation had won. In many other cases, shop owners prefer to close their factories during the elections for union representatives, and to reopen later with new workers and under the auspices of the official union.

Moreover, from 1987 the Nineteenth of September Union has faced tremendous problems in recruitment. By 1992 the official membership had shrunk to just under 800. Though the union could mobilise support from many more people this has reduced its financial capacity and bargaining position. The union has been powerless against declining wage rates. While the economic crisis has made some women workers more combative it has seriously limited their organisational participation. There are several reasons for the union's decline. The economic policy of the government has drastically reduced the buying power of workers, thereby limiting the time available to contribute to developing the union, as women must take on extra activities to bolster their family resources.

Another difficulty has been that some garment workers held false expectations. They thought that upon entering the union their working conditions would be transformed automatically. Because they were not used to active participation in a union they had tremendous hopes about simply joining and became correspondingly disillusioned when they did not achieve these desired changes.

Other factors blocking growth include: factory closures owing to recession and relocation outside Mexico DF; factory owners who threaten workers with dismissal if they become incorporated in the garment workers' union; a false characterisation of the union by factory owners and officials, who call it a 'red' union and claim that it has been responsible for closing some factories; the collaboration of factory owners and leaders with corrupt authorities which protect them if they act against the law.

So, while the general conditions of thousands of garment workers have benefited from the union's activities, new problems have arisen since it was formed.

Some of these were inherent in the system of production for, despite modernisation and industrial restructuring in Mexico, the garment industry did not develop like other sectors during the 1980s. The equipment in most workshops was not upgraded. Since there was a plentiful supply of labour, the workshop owners faced little incentive to modernise. Profits were secured by the intensification of labour. The manager would demand higher output during the nine-hour work day, reducing the operating costs per worker. Or the hours of work simply would grow longer through homework which women would do after they returned from the sweatshops. When recession forced many of the illegal workshops to close down, as the subcontracting chains to which they belonged collapsed and the local market contracted, the Nineteenth of September Union lost members.

Ironically too the union's very successes provoked retaliation which undermined its position. The union's defence of the workers led some factory owners to close or relocate factories. Others entered into labour conflict (including using violence as a means of defence) and dismissed union leaders rather than permit the union to legally represent the women who worked in their factories. Certain factory owners would not allow their relationship with workers to be mediated through a union but were resolved to continue in the traditional style, retaining absolute control to determine the economic incentives or rewards of their work-forces in a paternalistic manner rather than as a legal obligation.

Since the official unions operate under a corporatist system in Mexico, it is essential for them to maintain a large number of members, for this permits them to negotiate political posts which the government assigns. The loss of control of a factory reduces the force of these official unions, and they too have not hesitated to turn to physical violence, threats and punishments against women workers who question their conditions of work through the Nineteenth of September Union.

The methods of struggle which women garment workers are able to use have varied depending upon the strength of the movement and of worker–management relations; their tactics range from strikes and work stoppages to 'go slow' action. However, the lack of any law effectively preventing owners from closing factories to avoid any threat to their interests, and the existence of high unemployment, are major obstacles.

Despite the coercion and manipulation which the Nineteenth of September Union has experienced from the official unions and the deterioration of working conditions, it has maintained an alternative form of organisation for women workers in the garment industry, and has served as a point of reference for other sectors of women workers, which has helped to strengthen the workers' movement. Also the union has succeeded in gaining the right to negotiate collective work contracts for the workers who are affiliated with the union, even when their contract belongs to an official company union.

Other achievements have been financial and social. The union has helped workers to obtain loans and instituted a collective savings scheme, with the Movimento Popular, to enable workers to buy houses. The majority of workers count upon the factory to provide social security, holidays, a Christmas box of goods (although not always what has been stipulated under the law) and maternity leave. However, women homeworkers or women who work in clandestine workshops do not receive such benefits – and although there are no precise data it is certain that these women workers number several thousand. Using its limited resources, the union has begun to tackle their social needs: for example, it has created creches. The organisation has emphasised in its policies the specific problems which women suffer as women – issues which have previously been neglected. This perspective has extended the scope of unionisation. In addition to their labour rights the women garment workers have demanded more humane treatment, the installation of public services to alleviate the burden of work conducted in the home and revised laws which permit their development as workers and facilities to help in their activities as mothers.

In response to the changed economic and political context of employment, the garment workers made the following demands in 1990:

1 Increased wages and their prompt payment.
2 Access to creches supported by the state for all children of garment workers.
3 Workers must be signed up for social security at the beginning of the work contract.
4 Reduction of the work day, to conform with that specified by law.
5 Improved conditions of hygiene and safety in the workplace, notably regarding lighting, ventilation, cleanliness, toilet paper in the lavatories, clean drinking water and sufficient space.
6 Better treatment of workers by the shop owner and staff.
7 Overtime to be paid according to the law.
8 That sexual harassment against workers be eliminated when they decide to organise themselves.
9 Elimination of the practice by which women must prove they are not pregnant as a precondition for employment.
10 Cessation of the dismissal of women workers when they become pregnant.
11 An end to the abuses in the workplace and too heavy work which has made some women miscarry.
12 The right to free maternity leave.
13 Respect for the personal lives of women workers, since workshop owners already use personal arguments as a justification to dismiss workers or to deny them benefits.
14 Freedom for women workers to join the union of their choice.
15 Freedom for women workers to elect representatives of the union which they belong to.
16 Respect for collective work contracts.
17 Elimination of the 'protection contracts' for shop owners which are used to control and exploit women workers.
18 An end to the utilisation by shop owners of 'black lists' to discriminate against workers.
19 Abolition of the punishments, dismissal and harassment of women workers who organise themselves.
20 Commitment by the labour authorities, through an agreement to respect the rights gained to negotiate the collective work contracts.
21 The establishment of a salary contract law for all women garment workers, to be at least the minimum stipulated by law.
22 Employment security.

CONCLUSIONS

Women are a significant proportion among the millions in Latin America living at the extreme margin of poverty. The countries of this region are witnessing deteriorating conditions: economic policies during the 1980s, the 'lost decade', were insufficient to sustain and support their peoples. The role of women has turned into one of crisis management, for an increasing number have turned to any activity in order to earn a living. Thus, women have an increasingly large role in the informal economy, and they enjoy no labour security or benefits. The increasing number of women in production and commerce has not led to greater independence. The range of activities in which women are engaged means there is little time available for them to organise themselves.

State policies have had little impact in Mexico. A National Solidarity Programme (PRONASOL) has been introduced whose principal objective is to provide incentives to marginalised communities to develop their own projects for economic survival, with the support of the state for technical as well as financial assistance. PRONASOL has established a programme for 'women in solidarity', which aims to set up co-operatives in certain types of production so that women can produce and commercialise their own products. In addition to having very little financial resources, these programmes are limited and have mainly promoted activities traditionally associated with 'women's work', such as sewing, food preparation, nursing. This approach only reinforces the traditional role of women without a clear concept of promoting a sense of self-worth. Nor does it take into account how women's activities in the community or family can be integrated into policy. Thus, although public policy now aims to draw women into the productive sphere, their subordination continues to be reproduced.

In contrast, the survival of the Nineteenth of September Union is important as it presents an alternative organisation for women workers in the garment industry which is both independent and officially recognised. The union also serves as a reference point for women workers in other sectors, offering a model of new tactics to improve not only the conditions of work but also a means to make labour more efficient and to increase productivity in the industrial sector.

As an independent and democratic workers' organisation, it aims to convince diverse sectors of society of the necessity to gain the rights which are internationally accepted as basic; for example, the right to a decent salary; the right to receive technical training; the right to housing; the right to an education; the right to a health service; the right to free organisation and assembly; the right to have their children cared for in a creche.

Throughout its five years of existence, the garment workers' union has relied not only upon the effort, tenacity and patience of its members: its achievements would not have been possible without the ideological legitimation and economic support received from the national and international community. The moral support nationally and internationally and the recognition by numerous civil liberties and human rights organisations has confirmed that the struggle of the garment workers is just. This has helped to defend the organisation against the constant attacks by the garment industry owners and the official union organisation. The monthly fees that the union receives from its members are deducted from some of the lowest wages in the country. Thus, the total received from members accounts for less than 10 per cent of operating costs and permanent activities. The independent character of the Nineteenth of September Union in relation to the large official unions prevents it from using the economic resources which have been accumulated over many decades.

Apart from low wages and lack of resources, the key problem the union faces is its members' lack of time. In contrast to other industrial sectors, the owners in the clothing industry refuse to allow free time or paid leave for those women workers who hold a post in the union to carry out their union responsibilities.

Consequently the women garment workers face unique challenges in creating an organisational basis through a women's union. They need time and a space in which to reflect upon the problems of women workers in the clothing industry and they require opportunities to develop educational materials for women workers in the garment industry who are still ignorant of their rights. They also have to strengthen the political and union rights of those workers affiliated to the Nineteenth of September Union.

All of these objectives will be hard to realise given the current economic conditions of workers in Mexico. But while the Nineteenth of September Union has seen the number of its affiliates decline, it continues struggling. More than ever before, it needs help from supporters internationally if this unique example of an independent union oriented to women's needs is to survive and extend its influence.

Chapter 5

Self-Employed Women's Association

Organising women by struggle and development

Renana Jhabvala

ABSTRACT

For the great majority of working women in India, conventional forms of trades unionism are not possible. They labour sewing in the home, collect waste paper in the streets, are employed as building workers on contracts, or eke out a livelihood as small vendors. In the last two decades, partly inspired by the women's movement, grass-roots organisations have sprung up in India on an impressive scale. These have adapted themselves to the actual circumstances of poor women workers and devised new and creative means of mobilisation and defence.

The Self-Employed Women's Association (SEWA) is a notable example. It has successfully organised thousands of women in Ahmedabad, Gujarat. Its secretary, Renana Jhabvala, describes its origins and the manner in which it operates, both at the grass-roots and as a pressure group in relation to the local and national state as well as upon international bodies. SEWA's method is based on attention to the details of daily existence combined with a broader vision. Renana Jhabvala explains its mix of pragmatic persistence and aspiration for a co-operative economy; its capacity to defend as a trade union and welfare association while reaching out to new forms of work organisation through co-operatives.

The Self-Employed Women's Association (SEWA) was born in 1972 as a trade union of self-employed women. It grew out of the Textile Labour Association (TLA), India's oldest and largest union of textile workers founded in 1920 by a woman, Anasuya Sarabhai. The inspiration for the union came from Mahatma Gandhi, who led a successful strike of textile workers in 1917. He believed in creating positive organised strength by awakening workers' consciousness. By developing unity as well as personality, a worker should be able to hold his or her own against tyranny from employers or the state. To develop this strength, he believed that a union should cover all aspects of workers' lives, both in the factory and at home.

Against this background of active involvement in industrial relations, social work, and local state and national politics, the ideological base provided by

Mahatma Gandhi and the feminist seeds planted by Anasuya Sarabhai led to the creation by the TLA of their Women's Wing in 1954. Its original purpose was to assist women belonging to households of mill workers, and its work was focused largely on training and welfare activities. By 1968, classes in sewing, knitting, embroidery, spinning, press composition, typing and stenography were established in centres throughout the city for the wives and daughters of mill workers.

The scope of its activities expanded in the early 1970s when a survey was conducted to probe complaints by women tailors of exploitation of women workers and revealed the large numbers untouched by unionisation, government legislation, and policies. After this the TLA began organising self-employed women including stitchers, head loaders and vendors. In a meeting held to decide the future the women felt that a new organisation should establish itself as a trade union. This was a fairly novel idea, because the self-employed have no real history of organising. The new organisation was called the Self-Employed Women's Association (SEWA) and the first struggle SEWA undertook was obtaining official recognition as a trade union. The Labour Department refused to register SEWA because they felt that since there was no recognised employer, the workers would have no one to struggle against. We argued that a union was not necessarily against an employer, but was for the unity of the workers. Finally, SEWA was registered as a trade union in April 1972.

SEWA grew continuously from 1972, increasing its membership and including more and more different occupations within its fold. The beginning of the United Nations Women's Decade in 1975 gave a boost to the growth of SEWA, placing it within the women's movement. In 1977 SEWA's General Secretary, Ela Bhatt, was awarded the prestigious Ramon Magsaysay Award and this brought international recognition to SEWA. By 1981, relations between SEWA and the TLA had deteriorated. The TLA did not appreciate an assertive women's group in its midst. Also, the interests of the TLA, representing workers of the organised sector, often came into conflict with the demands of SEWA, representing unorganised women workers. The conflict came to a head in 1981 during the anti-reservation riots when members of higher castes attacked the *harijans*, the former untouchables. Many of the women were members of both the TLA and SEWA. SEWA spoke out in defence of the *harijans*, whereas the TLA remained silent. Because of this outspokenness the TLA expelled SEWA. After the separation from the TLA, SEWA grew even faster and started new initiatives. In particular, the growth of many new co-operatives, a more militant trade union and many supportive services have given SEWA a new shape and direction.

SEWA has grown with the inspiration and support of these three separate movements, and sees itself as part of a new movement of the self-employed which has arisen from the merging of all three. SEWA was born in the labour movement with the idea that the self-employed, like salaried employees, have rights to fair wages, decent working conditions and protective labour laws. They deserve recognition as a legitimate group of workers with status, dignity and the right to organise bodies to represent their interests publicly. Most importantly, the bulk of workers in India are self-employed and if unions are to be truly responsive to labour in the Indian context, then they must organise them. This requires going beyond the western model of a trade union as practised in industrially developed countries where labour is composed mainly of wage earners working for large-scale manufacturers or enterprises. In India, where only 11 per cent of the labour force is comprised of these types of workers, the trade unions must expand their efforts to represent the millions upon millions of self-employed landless labourers, small farmers, sellers, producers and service workers. Moreover, if unions are to be responsive to women workers, they must recognise that they are most concentrated in this sector. If labour unions want to touch the mass of workers in India and other developing countries, especially women workers, it is essential for them to organise the self-employed.

In addition, SEWA feels the co-operative movement is very important for the self-employed. Not only is it important for the self-employed to struggle for their rights, but they also need to develop alternative economic systems. The co-operative movement points the way to such a system: the workers themselves would control their own means of production in an alternative system where there is no employer and no employee but all own what they produce. Unfortunately, the world-wide co-operative movement has not really reached the poor. Workers' co-operatives have rarely been successful and co-operatives have been unable to change social and economic relations.

SEWA accepts the co-operative principles and sees itself as part of the co-operative movement attempting to extend these principles to the poorest women. In the present situation of our society, the co-operative movement has yet to reach poor women because the co-operative structure itself has been misused. The poor are consciously and deliberately excluded from membership. Women are not even perceived as part of the clientele, let alone as valid members in their own right.

The government itself has also weakened co-operative structures by intervening to control decision making, destroying the autonomy and voluntary nature of the co-operatives. SEWA sees the need to bring poor women into workers' co-operatives. The co-operative structure has to be revitalised if they are to become truly workers' organisations, and thereby mobilise the strength of the co-operative movement in the task of organising and strengthening poor women.

The women's movement in India began with the social and religious movements in the late nineteenth century. This period of seeing women as objects of social reform changed with the onset of the nationalist movement when, under Mahatma Gandhi, women actively participated in the freedom struggle and became active in their own liberation. In the 1970s, the women's movement took a new and more radical turn, with women participating actively in social movements and demanding opportunity in all spheres of life. The women's movement pointed out that women constitute 50 per cent of the world's population and they do two-thirds of all the work in the economy. For this work, women are paid only 10 per cent of all wages, salaries and remuneration. At most, 1 per cent of this income is owned by women. All this because women's work is not recognised as work, hence not paid or paid at very low rates.

SEWA has been a part of the growing women's movement. As the bulk of women in India are poor, self-employed and mainly rural, in order for the movement to be successful it must reach out to them and make their issues – economic, social and political – the issues of the movement.

ON ORGANISING

The basis for any movement has to be organising. And so SEWA's first priority is organising self-employed women. The self-employed woman is extremely vulnerable. She has no resources to fall back upon, no support structure, she is the weakest and most vulnerable person in our society, crushed under intolerable burdens. Her economic existence is precarious as she earns her living from day to day. She needs work desperately, but must compete with countless others like herself for dwindling work opportunities. Very often she has no work and so nothing to eat. Even when she works she earns less than anyone else in the economy.

Her work is hard and physically exhausting, often dangerous, and her hours of work are long. Her body is weak from ill health and unsafe childbearing. In addition, there is physical and emotional drudgery in the home. She squeezes as much as she can out of the meagre resources to feed her family; she gives emotional and physical care to the children, the old and sick. She is the last to sleep, the first to awaken; she keeps working in spite of illness, till she can no longer stand.

Her social existence is oppressive. She is usually of a low caste. She is a woman and so considered the lowest, and can be unwanted within the family. She is a worker without assets and so dependent for her livelihood on the powerful who keep her subjugated.

She is illiterate, she has no access to the resources of society, to education, to health care, to social security benefits. She is politically invisible, she cannot make her needs known to the politicians who take her vote. Her life is controlled by decisions made by others. Within her family her will is subject to the decisions of her husband, father, son. In her work she is controlled by her employer, contractor, landlord. In her social life she is bound by the rules laid down by her caste, *panchayat* or community elders in which she has no say. She is treated as a nonentity, a non-person, subject always to the needs of others.

These socio-economic conditions reveal an extraordinary vulnerability. Economically she is vulnerable, afraid of losing her work in a labour surplus economy if she asks for any more than she gets. Physically she is vulnerable to sexual attacks, to illness, to overwork; socially she is vulnerable to caste and patriarchal oppression. And underlying all this is the hidden but ever-present threat of violence.

This vulnerability makes it very difficult for her to organise. But the extreme pressures have in some ways made such a woman strong. She is able to survive under such crushing conditions only because of her deep faith, her courage, her love for her family and her indomitable will. She is weak, but her weakness is due to the pressures of society. She is weak as a social being in her relations to others, as a political being and in her social status. However, as a person she is strong, for her very social weakness requires that she be strong internally. In order to survive in a desperate struggle as the weakest in society she must develop internal resources of courage and strength. It is these strengths she draws on in the rare cases when she tries to fight back, to organise.

Over the last two decades SEWA has been trying to develop a joint strategy of struggle and development, where such women organise, both to form an alternative, and to combat their everyday oppression. SEWA's members are self-employed women. The list of trades is long, the examples of self-employed workers are numerous. Vast numbers of economically active women are the invisible workers of the nation, and also of the world. They rarely own capital or tools or production, they have no direct link with organised industry and services, and they have no access to modern technology or facilities. All they possess are the skills and knowledge of their trade and their physical labour. They constitute the majority of the enormous population of self-employed workers, normally called the 'unorganised sector'. In India only 6 per cent of working women are in organised industry and services; the remaining 94 per cent are left to fend for themselves.

These 94 per cent of the workforce constitute the self-employed workers of our economy. Self-employment is the major form of livelihood and includes all those people who have to earn their living without being in a regular and salaried job. There are three broad categories of self-employed workers:

1 Small-scale vendors, small traders and hawkers, selling goods such as vegetables, fruit, fish, eggs and other staples, household goods, garments and similar types of products.
2 Home-based producers such as weavers, potters, *bidi* (local cigarette) makers, milk producers, garment stitchers, processors of agricultural produce and handicrafts producers.
3 Labourers selling their services or their labour, including agricultural labourers, construction workers, contract labourers, hand-cart pullers, head loaders, *dhobis* (workers who wash clothes), cooks, cleaners and other providers of services.

The limited amount of data makes it difficult to assess the size and composition of self-employed workers. However, over the past several years, SEWA has conducted a series of socio-economic surveys of self-employed women workers which reveal a general profile of these women.

These profiles show that self-employed women are among the poorest of workers. Most of them are illiterate. Many of them have total family incomes of less than Rs 3,600 per year which is considered to be the poverty line cut-off (see Table 5.1). In addition, women workers make a very significant contribution to the total family income. In one sample study of workers (see Table 5.2) the income of more than 50 per cent of the women surveyed accounted for up to 50 per cent of the total family income.

Table 5.1 Socio-economic profile, 1990

Women	*Illiterate*	*Slum dweller*	*Married*	*Her monthly income*	*Family income*	*Place of work*	*Work outside*	*Take children to work site*	*Rented means of labour*	*Indebtedness*
	%	%	%	Rs	Rs	%	%	%	%	%
Agricultural labourer	90	–	90	200	3000*	–	100	–	–	–
Garment maker	57	55	72	100	352	99	1	–	20	67
Used garment dealer	70	80	83	200	300	–	100	65	–	25
Hand-cart puller	74	80	88	300	400	–	100	85	65	80
Vegetable vendor	90	95	86	300	400	–	100	39	49	80
Junksmith	96	70	90	300	400	99	1	–	–	10
Milk producer	90	30	80	193	374	100	–	–	–	35
Cotton-pod sheller	58	24	73	175	1800*	89	11	60	–	55
Handloom weaver	77	–	63	62	206	100	–	–	–	71
Waste picker	92	90	76	82	280	–	100	–	–	80
Firewood picker	97	–	75	120	200	–	100	–	–	42
Block printer	65	80	60	150	450	80	20	–	96	–
Bidi worker	70	89	71	250	–	100	–	–	–	64
Incense stick roller	67	80	80	201	589	100	–	–	–	32
Papad roller	50	82	54	300	500	100	–	–	–	20
Head loader	91	96	75	208	373	–	100	81	–	61
Bamboo worker	90	60	71	150	361	100	–	–	–	67

*(Seasonal)

Table 5.2 Distribution of respondent income as a percentage of family income, 1990

Type of work	*Up to 100%*	*11–50%*	*51–90%*	*91–100%*	*Unspecified*	*Total*
Agricultural labourers	8	72	9	27	13	129
Animal husbandry	5	10	2	11	1	29
Horticulture	0	2	0	1	0	3
Scrap collection	0	0	2	0	1	3
Hawkers and vendors	11	61	14	57	11	154
Home-based workers	19	140	8	32	6	205
Contract labourers in industry	6	21	0	4	13	44
Services	0	3	0	1	2	6
Transport	2	32	12	3	2	51
Other	5	5	3	6	9	28
Total	56	346	50	142	58	652

Two case studies give an insight into the complex and interconnected problems faced by poor self-employed women workers and show how organising is a continuous process. Detailed knowledge of the specific issues each group faces forms the basis of SEWA's strategic approach.

Naniben

Naniben is an agricultural labourer. She is about 40 years old but wiry and energetic. She says: 'SEWA has been working in our village for years. I was also a member and was part of the Milk Co-operative that SEWA had helped us form. Our village, Baldana, is so dry, there is no water, no trees, no vegetation. And during the drought (1984–7) whatever was there was stripped by the hungry cattle. Yes, I have one cow and when she gives milk I do earn something from the co-operative. But I have to spend so much time getting fodder. Usually, I give her to be grazed to the *bharvads* (shepherds) and I have to give them a share of the milk. Firewood is my other problem, my small son and daughter collect the fuel from the trees and bushes. I send them out in the morning and they come back sometimes two hours later, sometimes four or five. During the harvesting season it is easier because we can take the dry stalks where we work.

'Don't ask me about water. Our village well is dried up. Of course, the richer castes have water, because they have their own tube wells. But we take water from the pond. When there is no water in the pond, in the summer season, then the *panchayet* gets a tanker. Often it doesn't come. Then we go to the neighbouring village, 8 kilometres away.

'But our main problem is work. If there is work, we eat, if not we manage with some millet. If my children feel very hungry I beg for some food from neighbours. Then I put them to sleep. If you sleep you do not feel hungry, so I also try to sleep. You see, here there is not so much work. There are only two crops. So we can definitely get work for two months but after that we go to the brick fields in Dholka. But often we don't get work there either. My husband? He is also an agricultural labourer like me. He hurt his back last year trying to lift a heavy load of bricks. It was very painful, we took him to the hospital in Bavla. The doctor prescribed so many medicines but where to get the money from? So we had to borrow it, I don't have any land so I mortgaged some of my jewellery. It is my last piece – a very nice necklace that was mother's. But still he is not well and does not do much work.

'During the planting season we earn the most. Then we get two months work. We earn about Rs 250 for each acre to be planted with rice. Ten of us working for two days can plant an acre. But after monsoons we rarely get agricultural work. Sometimes we get weeding work but then we get only Rs 8 per day.

'Three years ago, we had a meeting with the SEWA organisers. We said we want to have agricultural work in our own village. That time, some officials had come from the government and they said, "You should plant trees." We said, "Yes, we also want trees because we get so many things from the trees – fodder, and fuel, fruits, medicines, toothbrushes (*dantan*). But we have no land to plant trees and no money for seeds and water."

'Champaben asked in that meeting "Can't we get work planting trees? We hear that some workers go to the forest department and they plant trees and are paid wages for it." Manjuben and Vandanaben [SEWA organisers] were there. They said, "Yes, that is a good idea. Let us ask the government for some land." We showed Manjuben that there is so much government land in our village but it is stony and dry and there is no water and no vegetables on it. "Even if we have this land what will we do with it?" we said. But Manjuben gave us hope. She said, "There are many ways of growing trees on this wasteland. And we can also dig a well later."

'So we approached the government to give us the wasteland. But they said, "First you must form a co-operative." We got the forms for forming a co-operative, but there were so many things to write and we are all illiterate. We felt, we can't do this, it is too much for us. But then, we had a meeting and Vandanaben asked us the questions and filled out the forms and we went to all the members' houses and took their signatures. So the co-operative was formed. Then, for one year we kept going to all the government offices and finally we also got the land.

'SEWA sent us to the Gujarat Agricultural University and we learned how to dig pits to preserve water as well as grow seedlings. We also learned how to make wastelands arable, which nitrogenous trees to plant, how to remove salts from the soil and many other things. When we got the land we had many meetings. The government officials came to our village and explained how payments would be made for working on the land [under the National Rural Employment Programme]. SEWA also brought some people to map out the land and we divided up the land among our members. It was summer and we began to work on the land, digging trenches and pits. It is very stony and hard land and it was very hot and the work was hard, but at least we were in our own village and could earn enough to eat.

Then, a big quarrel came: the *bharvads* became very angry. They said, "How can you people have this land? This is the land on which our cattle wander." They were angry because we had not asked them. They are very powerful in this village because they have so many households and they are very violent. They want their cattle to go wherever they like and then they claim the land is theirs. Also in our village the higher-caste groups, *koli patels* and *bharvads*, don't get along and they don't like *harijans* and *bhangis*. So they felt why should these lower-caste women get the land. We said we don't mind taking *bharvad* women in our co-operative but they don't like to do agricultural work like digging and planting.

'So the *bharvads* said, "We will not let the land go to these women." They threatened us: "If you go on the land we will beat you up." Then they filed a case in court and got a stay order. They were taunting us and saying "Where is your SEWA now? We are more powerful than any SEWA." We were very depressed and so were Manjuben and Vandanaben. All our efforts washed away! All our dreams gone! Even among ourselves quarrels began. Laxmiben's husband was threatened by Changanbhai Bharvad and she owed him a lot of money. So she came in the meeting and said, "We should not have done this, we should live peacefully in the village. It is better to have peace and eat one roti less." Some of the women were scared. I was also scared, but I said "Can't SEWA help us to fight the case?" So we got a good lawyer from Ahmedabad city and we fought the case. At every hearing we came to the Ahmedabad court. Such an experience! We sat in the front of the court where the judge could see us. Other sisters of SEWA joined us to make us feel strong. Sometimes the *bharvads* came too, they glared at us, but many sisters were there and we did not feel scared.

'Then we won the case. The judge said that government had given the land to Vanraji Co-operative and it is theirs. We were very happy. Next day we went to work on the land, but the *bharvads* stopped us. They said "If you go to the land you will not leave it alive." They stopped Manjuben and Vandanaben also and said they would kill them. The *bharvad* who stopped us has three murder cases pending against him and he has been externed from his village but still he lives there. So we did not go on the land.

'First the *sarpanch* was supporting us. But then there was a *panchayat* election and the *sarpanch* promised the *bharvads* that if he got released he would support them. So then, he began supporting them. So we had no support in the village. Some of us were still brave and we went ahead and planted the trees during monsoons. We kept a watchman to guard the saplings, but the *bharvads* sent their cattle to graze it all and the watchman was too scared to drive away the cattle.

'Finally, SEWA approached the *bharvads* through the Collector and the District leader of *bharvads*, and we had a compromise. We would only plant half the land and leave the other half for cattle. We all signed the agreement.

'Now we are working all right. We get work in our village and at the same time we are growing trees. Of course we had to give in to some extent but still we are better off than before. Our greatest strength is our unity.'

Karimabibi

'My name is Karimabibi and I am the vice president of SEWA. I live in Dariapur [a Muslim area in the inner city of Ahmedabad] and I sew quilt covers out of *chindi* [waste cloth pieces].

'My father had his own business of winding thread on bobbins which he used to get on contract from textile mills. We ate well and had nice clothes. Of course, my parents did not educate me much because in my time girls were not sent to school. I studied up to fifth class and at 11 years I left school and helped my parents to wind bobbins. I always liked arithmetic. I was married at the age of 15. My husband was a mill worker but he did not have good health and slowly we had to sell off everything; my jewellery, vessels, to pay for his medicines. I have three daughters, and when my youngest daughter was 7 years old my husband died and I was widowed, left with nothing but this small room.

'My daughters and I earned our living by sewing quilt covers. We would sew till twelve or one at night and out of that money I got my two daughters married. The youngest one refuses to get married and now we both sew quilt covers for a living. My brothers are well off, one of them is a *chindi* merchant, they help me sometimes but I do not like to spread my hands in front of them.

'I have always been active in fighting against oppression. I am always ready to fight even my own brother. I first joined SEWA like this: about twelve years ago, prices had gone up very high around Diwali, but the merchants were not willing to give us higher wages. So I said to all the women, let us go on strike. So we stopped stitching and we stood at the corners and would not let other women stitch either. But after a week our unity broke and we had to give up our strike. Women began stitching secretly at night. The traders did promise us an increase but they never paid it. We felt demoralised. But then one of our men said "Why don't you to go SEWA? They are a women's union and they will help you." He took some of us to Elaben [Ela Bhatt] and she sent Pallavi [a SEWA organiser] to do a survey. Then Elaben came and began organising all the women. We had meetings every evening and many women would come. When over 700 women became members of SEWA by paying a membership fee of RS 5 each, we wrote a demand letter to the traders. The traders tried to pressure us to stop going to the meetings. They called our men and said to them, "Your wife (or daughter, or sister) has become shameless. She goes out of the house and sits in meetings and shows her face to the world. Your honour will go down in the community." Because of this pressure some women stopped attending meetings. My brother who is a trader came to my house and said, "Have you lost all shame?" I said "You are my brother and I am your sister, that is our family life. But you are also a trader and I am a worker, so I will have to fight with you and other traders. There is no shame in coming out of the house for one's stomach. I am not doing anything wrong to have shame."

'SEWA complained to the labour commissioner and he sent inspectors to raid the traders' shops. Then he called the traders and us for negotiations. Seven of us went. This was the first time I had seen a government office. The traders said "We do not want to negotiate with these workers, we will negotiate only with Elaben." But Elaben insisted that we stay and the traders had to give in. So we sat across the table from the traders, instead of at their feet. After a lot of discussion they agreed to raise the rates to Rs 1 (from 0/60 paise) per quilt cover. The traders signed the agreement and so did we. I felt very proud to sign. We were feeling very happy.

'But really our troubles had just begun because the traders began victimising some of the women, who were widows and had no other support. We tried to help them by giving some work from our share, but how long could we do that? We are also poor.

'Again we called SEWA for a meeting: what would we do to help our sisters? I suggested we buy some *chindi* and sell the quilt covers ourselves. The Textile Labour Association [TLA] helped us, they gave us some money and we went to the mills and bought *chindi*. We distributed the *chindi* first in the TLA's sewing class and later they allowed us to use the library during the day. First we only gave work to the six women, but then so many women came saying "give us work also", and "we are needy". Some of them abused me because there was not enough work.

'From workers, overnight we had our own business. But so many problems. We went to the mills to get *chindi*. We were willing to pay money but the mills were not willing to sell it to us. Some mills sold their *chindi* only to the relatives of the owners. In other mills, especially government mills, the officers and clerks were getting cuts from the traders. We took the help of the TLA and met the higher officers. Then we get some *chindi* but the lower officers harassed us by filling bad *chindi*. In one *chindi* sack we got a shoe, in another a knife. Then we made friends with the people who sorted the *chindi* and filled the sacks. They were also poor women like us. After that we got much better *chindi*.

'But still getting enough *chindi* has always been a problem because the supply is limited. We tried to lobby the government and finally we got a concession – that we will get all the *chindi* from the government mills at 5 per cent concession rates. This has helped us a lot.

'Market was not so much of a problem because all the poor people in the villages use these quilt covers. But we have to keep the price low otherwise the workers could not buy, and with low prices we could not make much of a profit, and for some years we ran at a loss. Then we got a grant and bought our own shop and we were selling well from the shop. This year we have made a good surplus and all of us got vessels and money as bonuses.

'In the last eight years we have made a co-operative. SEWA called us one day for a meeting and said "SEWA is a trade union and can't run this business unit indefinitely so let us form a co-operative." At first we were scared, we thought SEWA would leave us and go away but then we were convinced that SEWA would help, so we called our co-operative Sabina – the name means belonging to us all.

'At first we could not understand – are we workers or are we owners? Are we employees or contractors? Because that is the only world we knew. Then we had a training class. We discussed the issue and we began to realise that in a co-operative we are both owners and workers which is something new. We are building something new and unique.

'In our area there is often communal tension. My house is in Nagina Pole just bordering Vadigam, a Hindu area. The trouble always starts in these areas. The Vadigam boys throw stones and burning sticks soaked in kerosene into our houses. Of course, our boys do the same.

'The worst riots were in 1985 and 1986. One boy near my house was killed, so his mother came to me, tears were running down her cheeks, and she said, "Please come with me to the morgue to collect his body." I went with her and we found that some people from Vadigam were sitting there too. They were waiting for a body also. We recognised each other of course because we are neighbours. At first we did not talk to each other. I was feeling very angry with them, but then we had to wait for four hours so we began to talk. Their boy had been killed in police firing and so had ours. We felt "Why is there a need to fight? We are being used by police and politicians." I said if there is any trouble in your area you call me and I will tell our boys to behave. After that there was not much trouble between Nagina Pole and Vadigam, but by then the trouble had spread to other parts of the city. We often had a curfew. We suffer so much under the curfew: there is no food, no milk for children, no work, no money, nothing. Our rooms are so small and all of us, men, boys, women, children, are all crowded together for days on end. Sometimes we feel it is better to die than to suffer these curfews.

'Then in 1986, although there was a curfew, the Rath Yatra [Hindu religious procession] came through Dariapur. They brought elephants and forced their way in; although army officers were there they could not stop them. The processionists attacked us, but the police fired on us and killed six of our boys. Then they clamped the curfew on us.

'We were all very angry, so all the women came out on the road. It was the month of Ramadan, we were fasting, it was very hot, but we said "We will sit here on the road day and night till you lift this curfew. You can shoot us and kill us but we will not move." There were thousands and thousands of women on the road. We did not let the men out of the houses for two days. We all just sat on the road. Finally the top army officer came and said, "What do you want? Send some of your men to talk to us." We said, "No, you talk to us women." He agreed to meet some of us, so the women told me, "You go from our side because you speak well and you are a vice president." So I went with three others to talk to the colonel. So he agreed to lift the curfew. But we insisted that he gave it to us in writing and he did. After that we all went back to our houses and there was no more curfew.

'But the riots spoiled the union we had made with so much difficulty. During the riots there were separate relief camps for Muslims and Hindus and our members were on both sides. So the unity we had built was broken by the riots.

'Before the riots we had tried to build our union. By then we had over 3,000 members not only of *chindi* workers but also other garment stitchers, especially petticoat stitchers. We went to the Labour Commissioner and said that we wanted minimum wages, but he said there was no minimum wage for garment workers. So we made an application to get minimum wages. But the file never moved. Later we found out that the big industrialists had sent many applications to stop the minimum wage. Their file was very thick but ours was thin, with only our one application. After the riots we found that our membership had dropped to less than 1,000 so we again tried to build up the union.

'We got very frustrated because the government was not bringing in minimum wages for us. So we took out a big procession with all our members. It was such a sight! So many women in *burqas* with little children. And we shouted, "We all are one," "Give us minimum wages." We went past the traders' shops and then we went to the Labour Commissioner's office and demonstrated there. Within the week the government took out the first notice.

'But then the industrialists began to send many objections. We also wrote many letters. We came to know that before being passed, the minimum wage has to be approved by the Advisory Committee. Some of the members are workers' representatives. So we went to them and explained all our problems to them and told them to pass the minimum wage. But the owners' representative got the meeting postponed three times, so six months passed. Finally the meeting was held. About 200 of us stood outside the meeting room. As the members came in we surrounded them and said, "Pass our minimum wage." The owners in the committee voted against the minimum wage but it was still passed. We were so happy. Some months later, the final notice came out.

'Then I went with our Secretary Renanaben to talk to the Labour Secretary about implementing the minimum wage. He said "Don't you know the minimum wage has been suspended? The Minister suspended it." Renanaben was so disappointed she started crying. But I said "No use crying; let us fight." So we all went to the Minister but he was very corrupt and had taken money from the industrialists and suspended the minimum wages. So we went to the newspapers and the whole story came out. We went and met the Chief Minister and the speaker. Then many of us went with leaflets during the assembly and gave leaflets to all the MLAs and told them to ask questions in the assembly. Then again we met the Minister and he brought back the minimum wage, so we won!

'Now we had the minimum wage but how to implement it? The traders were still not giving it. So we complained to the Labour Office and they made raids.

Then we went on strike and the traders also refused to give us work. The women were suffering because there was no work, but then the traders said to us, "We want to negotiate but only with you, we don't want the labour department." So they came to SEWA for negotiations. Fifteen traders came and there were eight of us in the team including Renanaben, myself and our garment workers executive committee members. The forty members of the trade committee were also there. They were like the audience. We negotiated for four hours. They agreed to raise the wages in three instalments but the main problem was thread. The women had to buy their own thread and when thread prices went up they lost money. We said the traders should give the thread. But the traders refused to do it. Finally we compromised and we said that our co-operative Sabina could open a thread shop at wholesale prices. The traders promised to contribute the capital for the shop. So Sabina opened the thread shop, but these traders haven't contributed anything as yet.'

COMBINING TRADE UNIONS AND CO-OPERATIVES

A movement needs organisational forms which will help it reach its goals and at the same time are in keeping with the goals; a means which will work towards the vision. The organisations to carry this movement forward must be capable of being controlled by the self-employed workers. They should be democratic and member-based. The trade union and the co-operatives are two organisations which can carry the movement forward. Both the trade union and the co-operative movement have a history of speaking for the weak and the labouring poor, of fostering a spirit of comradeship and of democracy. The goals of trade unions and co-operatives are the same but their methods are different. The trade union represents struggle, while the co-operative represents development. The trade union fights while the co-operatives build.

SEWA itself is registered as a union under the Trade Union Act, 1972, and is one large trade union whose membership comprises workers from many different trades and industries. Struggles of all kinds for earnings are carried out under the auspices of one union, also struggles for job security, better working conditions, social security, services, and changes of policy and law. Table 5.3 gives a list of trade union members.

Moreover, we have so far sponsored forty co-operatives. The first co-operative was the Mahila SEWA Sahakari Bank. Members of each trade are organised into different co-operatives. For example, the block printers co-operative, Abodana, is separate from the tree growers co-operative, Vanraji. The co-operatives are divided not only by trade but also by function.

Income and asset generation is carried out by producer and trade co-operatives whereas the provision of social security is carried out by service co-operatives. (A list of co-operatives is given in Table 5.4.)

Table 5.3 SEWA membership, 1990

Vendors		
Vegetable/fruit sellers	3,040	
Old clothes sellers	300	
Others	180	3,520
Tobacco workers and food processors		
Bidi workers	7,695	
Milk producers	3,158	
Tobacco workers		
Agriculture and processing	3,180	
Papad workers	80	14,113
Agricultural and allied workers		
Agricultural labourers	4,590	4,590
Textile and garment workers		
Quilt makers	680	
Garment stitchers	1,480	
Weavers	140	
Block printers	100	
Handicrafts	1,140	
Others	20	3,560
Other labour and service providers		
Paper pickers	1,400	
Head loaders	200	
Contract laboureres	400	
Others	240	2,240
Other Home-based workers		
Carpenters	20	
Household workers	260	
Childcare workers	40	
Agrabatti workers	180	
Others	630	1,130
Grand Total		29,153

Table 5.4 Co-operatives sponsored by SEWA, 1990

	Bank	
1.	Shri Mahila Sewa Shakari Bank	
	Livestock	
2.	Devdholera Dudh Utpadak Sahakari Mandli Ltd	
3.	Baldana Dudh Utpadak Sahakari Mandli Ltd	
4.	Lagdana Dudh Utpadak Sahakari Mandli Ltd	Ahmedabad
5.	Dumali Dudh Utapadak Sahakari Mandli Ltd	Dholka
6.	Ranesar Dudh Utpadak Sahakari Mandli Ltd	
7.	Rupal Dudh Utpadak Sahakari Mandli Ltd	
8.	Kathwada Dudh Utpadak Sahakari Mandli Ltd	
9.	Pasunj Dudh Utpadak Sahakari Mandli Ltd	Dascoi
10.	Vanch Dudh Utpadak Sahakari Mandli Ltd	
11.	Bilasiya Dudh Utpadak Sahakari Mandli Ltd	
12.	Miroli Dudh Utpadak Sahakari MAndli Ltd	
13.	Bahiyal Dudh Utpadak Sahakari Mandli Ltd	
14.	Amarajinamuvade Dudh Utpadak Sahakari Mandli Ltd	Dhegam
15.	Nanimorali Dudh Utpadak Sahakari Mandli Ltd	
16.	Waghpura Dudh Utpadak Sahakari Mandli Ltd	
17.	Navapura Dudh Utpadak Sahakari Mandli Ltd	
18.	Gulabpura Dudh Utpadak Sahakari Mandli Ltd	Radhapur
19.	Latia Mahila Dudh Utpadak Sahakari Mandli Ltd	
20.	Sherganj Mahila Dudh Utpadak Sahakari Mandli Ltd	
21.	Najupura Mahila Dudh Utpadak Sahakari Mandli Ltd	
22.	Kalyanpura Mahila Dudh Utpadak Sahakari Mandli Ltd	
23.	Gadh Mahila Dudh Utpadak Sahakari Mandli Ltd	
	Land-based	
24.	Shri Vanraji Mahila Sewa Vrukh Utpadak Sahakari Mandli Ltd – Dumeli	
25.	Shri Vanraji Mahila Sewa Vrukh Utpadak Sahakari Mandli Ltd – Metaal	
26.	Shri Vanraji Mahila Sewa Vrukh Utpadak Sahakari Mandli Ltd – Beldan	
27.	Shri Ganeshpura Mahila Vanlaxmi Sahakari Mandli (proposed)	
	Craft	
28.	Shri Sabina Mahila Chindi Utpadak Sahakari Mandli Ltd (Patchwork)	
29.	Shri Aabodana Mahila Sewa Chaapkaam Utpadak Sahakari Mandli Ltd (Block printing)	

Table 5.4 cntd

30.	Shri Bansari Mahila Sewa Vanskam Utpadak Sahakari Mandli Ltd (Bamboo work)
31.	Shri Vijay Vankar Mahila Utpadak Sahakari Mandli Ltd (Weaving)
32.	Shri Utsah Heatshall Oon Vanat Vankar Sahakari Mandli Ltd (Weaving)
	Trading
33.	Shri Saundariya Safai Mahila Sewa Utpadak Sahakari Mandli Ltd (Cleaning)
34.	Shri Shakfal Mahila Sewa Sahakari Mandli Ltd (Vegetables/fruits)
35.	Shri Matsyagandha Mahila Sewa Sahakari Mandli Ltd (Fish)
36.	Shri Jwala Kerosene-Vendors Co-operative
37.	Shri Swashriya Mahila Lok Swasthya Sewa Sahakari Mandli Ltd (Health care)
38.	Shri Sangini Mahila Sewa Sahakari Mandli Ltd
39.	Shri Pethapur Kagad Utpadak Sahakari Mandli Ltd
40.	Shri Sujata Mahila Sewa Sahakari Mandli Ltd

The long-term goals of both co-operatives and trade unions are to:

- increase income;
- provide assets;
- provide security of work;
- provide access to the social security services of health, childcare and housing and access to developmental services such as training, communications and banking;
- build solidarity and co-operation among workers;
- strengthen democracy;
- bring the self-employed women into the mainstream of national life;
- be equal partners and equal beneficiaries in the process of economic development.

Although the long-term goals of the trade unions and the co-operatives are the same, the short-term goals and the methods of operation are different. A trade union organises large masses of workers to act as a pressure group. It uses mass strength to voice the needs and demands of workers, to struggle and fight. In order to make the workers unafraid and courageous it uses various forms of collective action such as processions, *dharnas* (sit-down protests) and go-slows. The pressure of its mass membership enables the union to negotiate and bargain for better terms and conditions for its members. These strategies can be strengthened by the simultaneous use of the law to contest infringement of rights. The co-operative attempts to build an economic unit on the basis of sharing. It creates a spirit of co-operative sharing through one person one vote, equal ownership of assets, division of profits and co-operative education. It builds self-reliance through an alternative economic system, where worker and owner are merged into one. It brings the control of the economic system into the hands of the workers.

SEWA has been working on the joint strategy of trade unions and co-operatives since 1972 and our everyday experience continuously strengthens our belief in the effectiveness of combining both these forms. The co-operative helps the organisation of unions in several ways. One of the functions of a trade union is to struggle for higher wages or earnings. If the co-operative is in the same trade among the same set of workers as the trade union, the co-operative can set a standard and provide a model of higher wages or earnings. This allows the trade union to point out to the workers and the employers/traders that a higher wage/earning is indeed possible. It also gives the workers a leverage when bargaining with the employers/traders, local or national authorities.

One example from SEWA's experience illustrates this joint role. In the Taluka Dholka of Ahmedabad District many of the women are cattle owners who earn a part of their livelihood by selling milk. When SEWA started working in the area, the women were selling their milk to the milk traders at very low prices. With the help of the district dairy SEWA started organising the women into co-operatives and the co-operatives bought the milk at higher prices from their members. Hearing about these higher prices, the women in the villages in which there were no co-operatives began asking for and on many occasions getting higher prices from their traders.

Co-operatives also raised rates in sewing and embroidery. In Ahmedabad city SEWA organised the women sewing quilt covers out of waste cloth (*chindi*) into a union. Later SEWA started a co-operative, Sabina, in the same area. The co-operative was a small one of only 100 women out of a total 2,000 workers who were members of the trade union. However, since the co-operative paid a piece-rate almost double what the quilt cover merchants were paying, the trade union was able to demand and get an income of over 50 per cent.

The Banaskantha area is on the edge of the Rann (Desert) of Kutch and is one of the poorest areas in Gujarat with practically no employment available. But the women have a traditional art of beautiful embroidery. Some traders had been coming to the area and buying the embroidery pieces from the women at prices which earned them less than Rs 6 per day. SEWA started organising a co-operative (still to be formally registered) and was able to sell the embroidery pieces so that women made Rs 15–20 per day. The women then began demanding higher prices from the traders.

Another important aspect of information is helping victimised workers. One of the most difficult aspects of organising in a trade union is to overcome the workers' fear that they will lose their work. The co-operative can help the trade union to consolidate by giving work to victimised workers until they can be restored to their rightful place.

The Sabina Co-operative for example was started precisely because of victimisation during union action. When the *chindi* workers organised and asked for higher wages they did negotiate and get a better deal. But a few of the weakest of them were victimised by the traders and dismissed from employment. A fear psychosis set in among the other workers: they were not able to rally and began leaving the union. Under moral pressure SEWA began a small production unit to give work to victimised workers and this unit grew and turned into Sabina Co-operative. The co-operative gave the workers courage to stay with the union.

Co-operatives are also a vital means of creating alternative employment opportunities. The trade union strategy is often unsuccessful because ours is a labour surplus economy and there is just not enough work to go round. By creating additional, alternative employment in an area for a group of workers, the co-operatives increase the bargaining power of the workers. When SEWA first started working in the Dholka area, we found that wages for agricultural workers were extremely low. Agricultural work was available only for four months in the year, and the rest of the time there was practically no work. The need was for more work and SEWA helped to form co-operatives of handloom workers, spinners, potters and milk producers. The result was that the women had work all year round and in the agricultural season they were not desperate for work. They could bargain for higher wages. The wage went up 60 per cent in the area.

Co-operatives can be a source of knowledge about the industry. Organising the workers is only one aspect of trade union strategy. Another aspect is to understand the industry – its strong points, its weak points – and to use this understanding to get leverage for a better bargaining position. A trade union has contact only with the workers, so knows only about the aspects that the worker sees. However, the co-operative goes deep into the trade of any industry and so has a thorough knowledge of how the economic system works. It also knows about fluctuations in trade and changes in demand or prices that could be used for bargaining. This knowledge is invaluable to the trade union in formulating its strategy.

A good example is the cigarette (*bidi*) industry. SEWA has been organising *bidi* workers in Gujarat and Madhya Pradesh into a union. At the same time we have sponsored *bidi* workers' co-operatives in Jabalpur, Madhya Pradesh. In 1989 there was a crisis in the *bidi* industry due to a rise in prices and a shortage of the tendu leaf (in which the *bidi* is wrapped). The *bidi* manufacturers tried to pass on their losses to the workers by giving them low-quality and fewer tendu leaves and even enforcing wage cuts. However, because SEWA was deeply involved in the co-operative we knew the economics of the *bidi* trade. We refused to accept wage cuts and fewer raw materials, giving the manufacturer economic arguments based on our co-operatives' experience.

Sabina provides another instance of a co-operative contributing to an increase for other workers. During one budget year, the state government put a sale tax on *chindi* products, which were previously untaxed. The Sabina Co-operative products were also covered by the tax. However, the other merchants were very unhappy about the tax, and were lobbying to get it scrapped. Sabina came to know of this lobbying from the sales tax department and argued that unless the *chindi* workers employed by the merchants were paid higher wages, the tax should not be scrapped. The Labour Department intervened and wages went up once again.

Co-operatives can provide an alternative structure which can solve deadlocks in negotiations. One of the main functions of the trade union is to bargain for a better deal for its members. Often while bargaining the two sides get into positions which neither can let go. The co-operatives can then provide a *via media* or an alternative solution which is acceptable to both sides. For example, the home-based garment workers were agitating for a higher piece-rate from the merchants. Finally both sides came to the bargaining table. One of the points of dispute was that the workers had to buy thread and this was a rising cost for them. They demanded that the merchants buy the thread. The merchants had already experimented with buying thread and had found that it was too time-consuming and unwieldy for them. For some time it seemed that the talks would break down on this point. Finally a *via media* was found; the workers would buy the thread on a co-operative basis, the initial capital would be provided by the merchants and workers jointly, the workers would buy the thread as they needed it and every rise in the cost of thread would lead to an automatic increase in the piece-rate.

Trade unions have a 'hard' image. The strategy and tactics of a trade union lead to confrontation situations as the task of a trade union is to articulate conflicts and contradictions. The relationship of a trade union with other parts of society is therefore often non-harmonious. Furthermore the employers, government, merchants, traders, farmers, police or any other group with whom the conflict arises also deliberately promote an image of the union as unreasonable, destructive and violent. With this type of image, the union often loses potential allies such as the press, the middle class or any other part of society with which it has no conflict. This loss of allies and support can reduce the strategic options of the trade union.

When the co-operatives become part of the trade union it changes the image of the union and helps it win allies, sometimes even with the 'enemy'. An illustration of this occurred in the Kheda District, one of the richest districts in the country where the wealth is based on tobacco farming and processing. The tobacco farmers, merchants and processors are rich and powerful, controlling the economic, social and political life of the region. The agricultural and processing workers had been terrorised into accepting very low wages and miserable working conditions. Any attempts at unionisation had been put down with brutal violence and workers were scared. In this atmosphere SEWA had to be careful that our image did not become too hard or we would attract too much violence, lose support from the progressive elements and frighten away the workers. The trade union remained uncompromising in its demands for higher wages and better working conditions, but at the same time SEWA started work through health and childcare co-operatives in the areas. Both services were needed because of tobacco poisoning and our health and childcare work built a positive, helping image for us. Through this work we began to get the support of the *panchayats* and the district administration. Even the employers were shamed into supporting the childcare centres. Now the district administration backs us in our demands for higher wages; and as we have a talking relationship with the employers they are less likely to use violence against us and more likely to bargain.

Another example is that of *bidi* workers. Since both the employers and the government know that SEWA has a *bidi* workers co-operative as well as a union they are unable to build up a 'destructive' image for SEWA. Similarly, the *chindi* co-operative has established amicable, buyer–seller relationships with the *chindi* merchants, so when the union raises demands, the merchants do not get into a totally hostile framework.

The process is reciprocal: the trade union also helps the co-operative. The trade union has the capacity for mass mobilisation, pressuring tactics, fighting strategies, legal know-how and effective bargaining. And often the co-operatives need all these things. Trade union power can be exerted to change policies. Poor women's co-operatives work under many disadvantages. There are internal disadvantages, such as lack of know-how and lack of cohesion, and external disadvantages, such as attack by vested interests. A major disadvantage that poor peoples' and especially poor women's co-operatives work under is that the weight of the economic system is against them.

In order to run smoothly the co-operative needs policy decisions and changes; these can be brought about only with the pressure of trade unions' mass support. A successful case was the paper pickers. SEWA had been organising paper pickers into co-operatives. One of the co-operatives helped them to collect their paper and sell it at reasonable prices. The trade union helped the paper pickers to get a special government resolution that all waste-basket paper should be given free to the co-operative members. When the validity of this resolution expired the government did not renew it owing to pressure from paper traders who wanted the free paper for themselves. So the trade union organised a mass meeting of over 2,000 paper pickers, invited the Chief Minister to the meeting and presented him with their demands. Immediately, a new resolution was passed.

Milk producers also benefited from trade union support. Women's co-operatives were having problems selling their milk to the dairy as there was overproduction of milk and the dairy refused to buy it. The trade union organised the dairy co-operative members, especially the secretaries, to present their case, with the result that the dairy had a special policy of total milk purchase for women's co-operatives.

When the Sabina Co-operative was facing losses because of lack of raw materials, SEWA garment workers' unions pressurised the board members of the National Textile Corporation to pass a resolution giving Sabina all its raw materials at lower rates. Opposition from vested interests can threaten members of co-operatives as they consolidate their organisation; in these cases trade union solidarity can be decisive.

The Saundarya cleaners' co-operative faced an attack from a trade union in the organised sector. The hostile trade union attempted to break up the co-operative, turning its members against Saundarya. It tried to discredit the cleaners' co-operative by projecting Saundarya as a 'contractor'. SEWA protected Saundarya in the following ways: it initiated talks with the external trade union on a union-to-union basis and reached an agreement on mutual co-operation. It also mobilised Saundarya members to shun the other union. It enlisted public opinion against the external trade union and for Saundarya by working among the *harijans* who were Saundarya members and in the wider *harijan* community. Here the paper pickers who were SEWA members but not part of the cleaning teams were especially helpful. Links with the *harijan* community were accentuated by the fact that as the other union had no *harijan* support SEWA was able to gain support in the press for Saundarya and was successful in convincing the Labour Department in its favour. SEWA also worked through the law. The trade union legal cell fought the case on behalf of Saundarya in the Labour Court.

The oral accounts illustrate this joint strategy. The paper pickers' co-operative had a government resolution that it could get free waste paper from the government printing press. In spite of this, private traders, with the connivance of the clerks and watchmen, were illegally taking away paper. To prevent this the union mobilised the paper pickers to confront the traders. One night all the paper pickers assembled before the press and caught a trader actually carrying away paper in a truck. They lay down in front of the gates and refused to allow the truck to go till the police came.

Union action has also been crucial in rural areas. A major attack on Vanraji Tree Growers' Co-operative was launched by the *bharvad* community in a village in Dholka Taluka. Vanraji, composed of poor landless village women, had been given waste land to grow trees by the government. The powerful *bharvad* community was unhappy partly because it showed they had less power in the village and because they were losing ground for their cattle to graze. They therefore filed a case in the civil court and obtained a stay order on the transfer of land. The trade union helped Vanraji to fight and win the case in five separate courts. It also helped Vanraji to negotiate a final compromise solution with the *bharvads*.

The combination of trade union and co-operative power makes it possible not only to defend members but to present an ideological alternative. Poor women's co-operatives are a new phenomenon. SEWA has a vision of the co-operative as a form which will bring about more equal relationships and lead to a new type of society. This concept is often not understood, and the co-operative is treated like a private trader or contractor. The trade union helps to articulate and reinforce the co-operatives' identity. For instance, Sabina had to buy waste cloth from the textile mills. In order to do this the mills insisted that it fill a tender, like any other trader. SEWA objected that Sabina was a poor women's organisation, not a trader and so would not fill tenders. Eventually a system was worked out where Sabina could buy waste cloth without filling tenders. On another occasion when Saundarya was under attack from the rival trade union which tried to make out that it was not a poor people's organisation but a contractor, SEWA argued Saundarya's case in the Labour Court. We used the dispute to articulate Saundarya as an alternative co-operative concept. This definition was clearly written down and has become part of the legal documentation which is also to be considered in the High Court. SEWA's trade union also convinced the Labour Department that Saundarya is not a contractor and so does not need a licence under the Contract Labour Act. It convinced the Labour Department to write a letter stating this.

The co-operative's success depends on establishing a certain level of integration among its members. Co-operatives which are formed after trade union mobilisation are much quicker to acquire this cohesion as the workers already have a feeling of unity based on common struggle. The Sabina Co-operative for instance had relatively few internal tensions in its first two years as it was formed after trade union mobilisation. Similarly Vanraji (Baldana) has high morale and sense of unity as a result of its contest for land.

Interaction between a trade union and co-operatives can be a continuing process of mutual support. During the course of trade union organising issues emerge which are taken up by the co-operatives. And during the establishing of co-operatives issues emerge that are taken up by the union. An example of a trade union generalising problems which are taken up by co-operatives is the Sabina Co-operative. The need of the garment workers for a wholesale thread shop emerged during negotiations and bargaining with the employers. The formation of the shop was accepted by the employers. It led to the formation of the Sabina Co-operative. Another case occurred when the trade union was organising *bidi* workers in Madhya Pradesh and Gujarat. In Madhya Pradesh the *bidi* employers were moving away from the towns into the villages. For example in the Jabalpur district *bidi* production had moved from Jabalpur town into the surrounding villages and the *bidi* workers in the town were totally without work. The need was for more work and this was taken up by the co-operative formed through SEWA.

Equally there are examples of issues emerging from the co-operative as a form of organising which are taken up by the union. In Banskantha SEWA has been organising artisans into co-operatives to help them produce and market their products. During the course of the organising it was found that the Gujarat Handicrafts Corporation, which was founded by the government of Gujarat for the very purpose of the development of artisans, was in fact buying from the middleman and very little benefit was reaching the artisans. The trade union took up the issue on behalf of all artisans, had a meeting of artisans all over Gujarat, and passed a resolution with a charter of demands which were handed over to the Handicrafts Corporation. The trade union then continued to lobby till the corporation was more responsive to the problems of the artisan.

The combination of trade union and co-operative serves the all-round needs of the workers. This leads to their development in every field and, at the same time, they come to only one organisation for everything and hence feel more loyalty. Furthermore, when one activity is not doing well the workers will still stay with SEWA because they value the other activities.

SEWA maintains a balanced image. The trade union keeps it radical and the co-operative conservative. In the *bidi* trade, for example, we have knowledge about the production and marketing side through the co-operative and about workers' numbers and the spread of work relationships from the union. Any change in the economic arrangements is immediately known through either co-operative or union. SEWA similarly learns quickly of any change in government policy. Moreover, our input into planning goes through labour as well as industry policy-making channels.

An interesting relationship has developed with government officials. Usually when a new official comes in, it takes a while to establish contact based on trust and respect. Since most officials are transferable we find that when a relationship is established with an officer through one aspect of SEWA's work, these links can continue and be of advantage in other contexts. After we established a working relationship with the Labour Commissioner, he was transferred to the Gujarat Handicrafts Board, where we did not have to start from square one but could start dealing with him immediately.

WHAT NEXT?

SEWA's General Secretary Ela Bhatt often says that it takes five years of hard work to make a project successful. In ten years, if you go about it the correct way, you can build an organisation. To become a movement, if the conditions are right, requires twenty years of work. If Elaben's theory is correct then SEWA, which was founded in 1972, is entering a new phase – it is turning from an organisation into a movement.

The indications are apparent. SEWA grew fast in the 1980s. Our membership went from 6,000 in 1981 to 30,000 in 1990 in spite of two years of rioting and four years of drought; the present membership is now 46,000. One co-operative in 1981 had risen to forty co-operatives in 1990. The bank has grown from a working capital of Rs 3 million to Rs 15 million. In the last ten years, nine new SEWAs have started in five states of India. We have added many new areas – communications, childcare, housing, insurance, training and research, and land and water. We have achieved visibility at the political and policy levels. Our General Secretary has been a Member of Parliament and is presently a member of the Planning Commission. We demanded and got a National Commission on Self-Employed Women. Some of the issues we have taken up, legal protection for home-based workers, national policy on vendors, women's banks at district level, are receiving serious attention at national and international level. It is evident that a series of challenges faces us as we turn more and more into a movement.

The first challenge is that of growth. Our growth has to be continuous and balanced. The unions, the co-operatives, the supportive services, all have to grow at a pace to support each other. As SEWA gets better known, there is a demand to start new SEWAs in other parts of India. This outward growth has to occur at a steady pace yet without over-reaching ourselves.

The second issue is structural. How do we keep all the organisations within the SEWA movement? How do the organisations grow through their own activities and yet support each other and remain part of a common ideology and a common movement?

The third question is one of leadership. For both the movement and the organisations it is necessary to have leaders at the national, state and local levels. So far, SEWA has been successful because it has been able to attract women from both the working class and the middle class and help them develop into leaders at all these levels. If the movement is to succeed then we need to develop leadership especially in local organising. At present SEWA is confronting the issue of how to encourage working-class women to become leaders at the state and national level.

A further problem is moving into a more central role in the trade union and co-operative movements. In the last ten years both the labour and the co-operative movement have begun to recognise and listen more to the self-employed. However, the trade unions are dominated by formal-sector and especially white-collar workers, while the co-operative movement mainly represents the rich. Both are defined and controlled by men. The challenge is to move self-employed women into centralised roles and leadership in both these movements. This is linked to gaining a more centralised position in policy making and advocacy. A start has been made with the National Commission on Self-Employed Women and Elaben's role as a member of the Planning Commission, but we are in most ways still peripheral.

Another type of challenge involves initiating new areas of work successfully. These include: rural banking, getting land in women's names while developing it as a co-operative, the economic development of a district through women's leadership (now being tried in the Banaskantha District), health insurance and social security benefits through co-operatives, consolidating our educational work by creating our own 'university' and strengthening our alternative media of video and newspapers.

So really we have only just started. We have a long way to go.

GLOSSARY

bhangi	a caste or community of cleaners, hide-flayers, part of the caste lowest in the hierarchy
bharvad	shepherd
bidi	cigarette
burqa	the head-to-toe covering, usually black, sometimes white, worn by Muslim women over their clothes to cover themselves. It also covers their faces
chindi	waste cloth
dharna	sit-down protest
harijan	literally meaning 'Children of God', the name that Mahatma Gandhi gave for the former untouchables (they include *bhangis*)
panchayat	the elected village council which decides most of the civil adminstrative and often social matters of the village
sarpanch	the elected head of the *panchayat*

Chapter 6

Deindustrialisation and the growth of women's economic associations and networks in urban Tanzania

Aili Mari Tripp

ABSTRACT

The austerity measures which have brought Latin American women to act over prices in their communities and driven even desperate low-paid sweatshop workers to unionise have also severely affected urban African women. Women in Tanzania have been forming their own economic links and networks in an effort to create some means of livelihood amidst inflation and economic crisis. Their new role as important contributors to the urban household income has had an impact on gender relations and marks a shift away from the state to private resources and solutions to making a living.

Aili Mari Tripp has studied the Tanzanian economy and society extensively and was assisted by Salome Mjema in her research. She looks at both voluntary associations and a variety of economic and social networks which include rotating credit groups, dance groups, childcare networks, meetings at the hairdressers and the market. Organisation embedded in everyday life enables women to withstand crisis in the public sphere. They also reveal needs and priorities of women neglected in top-down state policies. The question raised by this study is: what aspects of these survival networks might form the basis for a viable and democratic development?

In many developing countries, one of the consequences of the world recession that started in the late 1970s and the economic restructuring programmes that followed has been an expansion of local women's organisations to cope with the new hardships they face as a result of the imposition of various austerity measures.[1] From the neighbourhood committees and communal kitchens of Peru, to the mothers' clubs of Brazil, and the 1985 demonstrations of women in Sudan against rising prices, women have been active in responding to the new pressures they face. Similarly in urban Tanzania, women's economic associations and networks increased in number and expanded in size in the 1980s as more women became involved in income-generating activities. Tanzania, like many African countries, was severely affected in the late 1970s by pressures from the world economy and drops in export commodity prices, which contributed to a dramatic decline in the real wages of workers. This crisis, followed by a series of economic reforms which had a negative impact on industry, pushed large numbers of women, in particular, into informal income-generating activities.

Urban women formed organisations as one of the strategies they pursued to deal with the new economic difficulties they faced. These associations emerged in a context where the ruling party and government had increasingly curtailed their activities after independence in 1961 and discouraged the growth of new autonomous organisations, especially economic ones. This meant that while some well-established associations like the Young Women's Christian Association (YWCA) continued in the face of various proscriptions, most independent women's organisations were confined to being small and informal in character.

By the early 1990s, however, the government and party began to accommodate such voluntary associations to a greater extent, in part because they did not have the capacity to curtail their activities, but also because there was an increasing realisation that a wider array of forces could be useful to the development process. The necessity for a greater plurality of organisations was reflected in the initiation of the debate in 1990 by former Party Chairman Julius Nyerere over whether Tanzania should move from a one-party system to a multiparty system.

This study of women's associations and networks starts by looking at the process of deindustrialisation as the consequence of the economic crisis and various economic reform measures. I show how it affected the urban household economy and, in particular, women, who increased their involvement in income-generating activities. I then explore the concomitant rise of women's associations and networks and the role they played in sustaining women's economic activities during this period. It is based on fieldwork carried out by an assistant, Salome Mjema, and myself between 1987 and 1988 in Dar es Salaam, Tanzania's largest city. The research involved hundreds of unstructured interviews and several surveys, including one cluster survey of 287 residents in Manzese and Buguruni, two parts of Dar es Salaam, made up primarily of workers and the self-employed. We also talked to women leaders and members of a variety of formal and informal organisations. I have also drawn on interviews conducted by Marja-Liisa Swantz during 1989.

CAUSES OF DEINDUSTRIALISATION

Like many African countries, Tanzania suffered in the 1970s from drops in export commodity prices; increases in import prices, especially of oil; and rising interest rates. Tanzania also suffered in this period from drought, from its war with Uganda and from the breakup of the East African Community, Tanzania's economic consortium with Uganda and Kenya.

These external causes of crisis were compounded by internal policies, including the heavy investment in industrial programmes at the expense of agriculture, in a country where 85 per cent of the population is engaged in agricultural production. Unfavourable government prices for producers of export crops resulted in shifts away from the production of cash crops to the production of food crops and to sales of both export and domestic crops on unofficial markets rather than through state-run crop authorities, thus undermining the state's main resource base.

The crisis that ensued took many different forms. The trade balance, which had been in surplus in the 1960s, had fallen into deficit by the 1970s. In the late 1970s, Tanzania experienced a severe and chronic foreign exchange crisis, which subsequently affected agricultural and industrial production, both of which are heavily dependent on imported inputs. In the 1960s Tanzania had the highest rate of increase in domestic food production of the entire African continent and was exporting grain to neighbouring countries. By the mid-1970s, Tanzania's agricultural production had declined to the point where it was forced to import food.[2] Industrial production had fallen to an annual –11 per cent growth rate by 1981.

The crisis limited the government's options significantly and forced it to initiate a series of economic reform programmes. The first two programmes which were initiated (the 1981 National Economic Stabilisation Programme and the 1982 Structural Adjustment Programme) were largely unsuccessful. Two Economic Recovery Programmes (ERP) followed in 1986 and 1989, including exchange rate adjustments; the raising of official producer prices; lifting of price controls; improving foreign exchange allocations; and efforts to raise the level of domestic savings, to improve the infrastructure and to launch major rehabilitation projects.

In spite of the economic reform programmes, Tanzania's foreign exchange crisis persisted, while pressures from foreign donors mounted, forcing Tanzania to negotiate an agreement with the International Monetary Fund (IMF) in 1986 after a six-year stalemate in talks. The IMF agreed to a standby arrangement subject to various criteria, which included substantial devaluations; restrictions on the amount of credit that could be transferred from the banking system to government institutions; limits on the accumulation of new debt; and controls on new external borrowing and on the overall budget deficit.

One of the main outcomes of the 1986 Economic Recovery Programme was a rise in growth rates, with the GDP increasing from –4 per cent in 1981 to 4 per cent in 1989.[3] Production of crops like coffee, tea, cashewnuts, tobacco and sisal increased in response to price incentives. However, the programme did little to improve the country's industrial production, which continued to register negative growth rates throughout most of the 1980s, although there was a slight improvement in industrial performance by the end of the decade.[4]

After 1985 both exports and imports grew, but the total import bill was still three times greater than export receipts in 1990. Even though export volumes grew in 1987, for example, export earnings were not much greater than the previous year because of declines in international commodity prices. By 1988 Tanzania's debt had risen to almost twice its GNP and ten times its export earnings. This is high compared to the rest of Africa, where the debt is equal to GNP and 3.5 times export earnings.

One of the most alarming consequences of the Economic Reform Programme, however, was the manner in which government expenditure was redistributed. Social, public and welfare services as a percentage of government expenditure had decreased substantially in the 1980s and by 1986 they had dropped to their lowest point in twenty years. The amount spent on health, education, housing and public, community and social services was cut in half between 1981 and 1986, from 21 to 11 per cent of total government expenditure. Attempts were made to reverse these cuts with the 1989 Economic Recovery Programme–Priority Social Action Programme.

CONSEQUENCES OF DEINDUSTRIALISATION

As mentioned previously, while the 1986 Economic Recovery Programme managed to reverse the decline in agricultural production, it did little to reverse the declining trends in large-scale industry. (Small-scale manufacturing, in contrast, experienced a boom in this period.) The crisis in industrial production could be largely attributed to foreign exchange shortages which affected industry's ability to purchase raw materials and machine spares and caused interruptions in the supply of infrastructural inputs. By 1987 Tanzanian industry was dependent on imports for 70 per cent of its inputs.[5] Although the investment structure had favoured the expansion of an already underutilised industry, with massive capital inflows at a rate of 20 per cent a year from 1978 to 1982 in monetary GDP, large-scale industrial production declined throughout the 1980s and by the late 1980s industry was operating at 20–30 per cent capacity.[6]

The 1984 trade liberalisation policy placed an even greater burden on various public and private industries, e.g., textiles, which found they could not compete with cheap imports. Textile production was especially hurt by the importation of *mitumba* (second-hand clothes) and better-quality materials.

In sum, the ERP adversely affected industrial performance in a number of ways. Tight monetary policies hindered industries from obtaining loans for financing working capital; trade liberalisation eroded the domestic market for various industries; devaluation increased the price of directly imported inputs; and raising producer prices stimulated the production of crops but resulted in sharp increases in the cost of inputs.[7]

URBAN TANZANIAN WOMEN'S RESPONSES TO DEINDUSTRIALISATION

For workers, this process of deindustrialisation meant sharply declining real wages and layoffs. There had been an 83 per cent decline in the real income of wage earners between 1974 and 1988.[8] Prices increased 5.5 times from 1982 to 1988, while wages only doubled in this same period. For a low-income household food expenditure alone exceeded the minimum monthly salary eight times. Because wage earnings could no longer support the household, the burden fell largely on women to sustain the household through informal small businesses. Because of low wages and discriminatory hiring practices, women were excluded from formal employment. As real incomes dropped for men, women were therefore more likely to initiate income-generating activities, frequently making them the main breadwinner in the family. In 1976, wages constituted 77 per cent of the total household income, while other private incomes accounted for only 8 per cent of the total household income.[9] In contrast, by 1988, according to my survey in Dar es Salaam, informal incomes constituted approximately 90 per cent of the household income, with wage earnings making up the remainder.

Thus, women, who in the past had contributed relatively little to the urban household income, were now critical to the very survival of the household through their involvement in projects. For low-income women these projects ranged from making and selling pastries, beer, paper bags, and crafts, to tailoring, hair braiding and even urban farming.

By the late 1980s it was women themselves who were often rejecting wage employment and seeking more beneficial incomes from self-employment. Women as a proportion of the employed workforce in Dar es Salaam had risen from 4 per cent in 1961 to 11 per cent in 1967 and 20 per cent in 1980. By 1984 the proportion had dropped to 19 per cent and if my survey is any indication it remained at 19 per cent up to 1987.[10] Between 1953 and 1980 few women left their jobs; according to my survey, however, the numbers more than doubled between 1980 and 1987. The most common reason both men and women (45 per cent) gave for leaving was low wages. The second most frequent response (17 per cent) was layoffs due to factory cutbacks or closures. Women were hit especially hard by these layoffs. For example, in 1985 when 12,760 workers were laid off in a retrenchment exercise, most of them were women.[11]

As one woman hairdresser who also sold charcoal and fried buns put it:

> 'In the past women took care of the children and didn't know how to make money for themselves. Now women have to find work [self-employment] because this is the way things are progressing in the country. You can't wait for your husband to give you money. You have to go out and find work for yourself. That is why you find women today doing all kinds of things like making *maandazi* [buns], selling *mitumba*, sewing, everything.'[12]

Women's changing role is also borne out by comparing various surveys. A 1970–1 study found 66 per cent of women in Dar es Salaam with no source of income; 13 per cent of the women were wage earners; and 7 per cent self-employed.[13] In contrast, my 1987–8 survey found 66 per cent of women self-employed while 9 per cent were wage-employed. Similarly, another survey of 134 women in Dar es Salaam in 1987 showed that 70 per cent of all women had projects while 5 per cent were employed.[14] Sixty-five per cent of those who started small businesses between 1982 and 1987 were women, while in the previous five years women made up only 28 per cent of those starting businesses. Of the self-employed women 67 per cent began their enterprises between 1980 and 1987, while only 15 per cent started in the 1970s. Women workers, like men, frequently had sideline businesses because wages were so low relative to the income they could obtain from a small enterprise.

Participation in the informal economy had resulted in massive reversals in dependencies and reversals in the direction of resource flows at many different levels: away from the state towards private solutions to problems of income, security and social welfare; away from reliance on wage labour to reliance on informal incomes and farming; and a gradual shifting of migration patterns with increasing new movements out of the city into the rural areas. Similarly, relations and patterns of obligation in the household were being transformed with greater resource dependencies on women, children and the elderly, where only a decade earlier urban women had mostly relied on men for income, children on their parents and the elderly on their adult children.

The pervasiveness of urban women's involvement in business was unprecedented in a country like Tanzania. This major change went relatively unnoticed by the government, scholarly and donor circles since much of it was informal and therefore unrecorded. Moreover, women's informal work was often considered unimportant and petty because of the bias towards official employ- ment. But gradually men were, at least privately, beginning to take note of these changes. As the head of the National Bank of Commerce told me in 1987: 'Women have become the largest private sector in Tanzania, but no one knows what they do.'[15]

One indication of men's acceptance of women's increased involvement in micro-enterprises was the fact that 44 per cent of women with projects had received starting capital from their husbands. Several men indicated their apprehension, especially about the fact that women were making so much more than they. One way for women to avoid open conflict with their husbands over their new access to cash was not to disclose their full earnings to their spouse. This was also a way of ensuring control over their earnings, which had to be stretched to cover the household expenses.

The new economic importance of women to the household was also changing women's perceptions of themselves. As one woman, who had left employment as a secretary at a state-owned company to start a hairdressing salon in 1984, explained:

> 'I started a business because I did not want to use my body to get money. I had four children to support and my salary was not enough. I wanted to use my own hands, brains and wanted to aim high. Whatever a man can do, I can do better. If you start aiming high, the sky is the limit. . . . Most women are in business and are no longer dependent on men.'[16]

WOMEN AND THE LABOUR MOVEMENT

Women found the national labour union, the Trade Union of Tanzanian Workers (JUWATA), of little use in protecting their jobs or fighting for increased wages. Moreover, women had little impact in the national labour movement, in part because they were not involved in the sectors of labour most active in the union. Although all workers in the formal sector automatically belonged to the union, women generally did not participate in JUWATA actively. It was widely acknowledged that when women were forced to attend meetings they simply remained silent.[17] In part, this could be attributed to the political culture in Tanzania, where women in mixed gatherings often found it improper to speak in public. However, a more plausible reason for this lack of involvement by women was the fact that they rarely found the union leadership interested in taking up their particular concerns. In fact, by the 1990s it is doubtful that many workers, male or female, found the organisation truly representing their interests.

Although the labour movement in the 1930s, 1940s and 1950s had vigorously fought and succeeded in keeping wages somewhat in line with the cost of living, by the 1990s it had virtually no autonomy from the ruling party and from company managers, who ran the union! Like many independent voluntary associations and independent organisations, the trade union movement was gradually co-opted after independence in 1961. Independent economic organisations like trade unions and co-operatives were seen as especially threatening to the party's control and government stability. By the 1980s, when workers did resort to independent strike action, they faced bullets and arrests. In 1986, for example, three workers were shot by a special police unit, seventeen injured and thirteen arrested at Kilombero II Sugar Company when 500 sugarcane cutters protested low wages and the exclusion of a customary bonus from their pay cheques.

Not surprisingly, JUWATA failed to respond to the dramatically declining real wages, layoffs and the other effects of deindustrialisation on labour. JUWATA leaders, in fact, expressed support for the Economic Recovery Programme because it included pay increases for workers and because policy makers had rejected IMF pressures to freeze wages in drawing up the ERP. But these pay increases were negligible when compared with the rising cost of living. Moreover, the ERP undermined workers' welfare in other ways, not only through its impact on deindustrialisation, but also through its cuts in social and public services.

Although women workers, like workers in general, had largely abandoned the union as a forum to promote their interests, this is not to say there was no opposition to government and party policies from labour or civil servants. There was in fact fierce resistance, in the form of quiet day-to-day non-compliance against policies which encroached on people's ability to pursue income-generating activities *outside* their formal jobs. Such resistance ultimately contributed to the greater legitimisation and legalisation of informal-sector activity and attempted to create a better environment for the pursuit of such small-scale private enterprises.

WOMEN AND THE NATIONAL WOMEN'S UNION

Just as women could expect little assistance from the labour movement in protecting their interests on the job, similarly the party-affiliated national women's union, Umoja wa Wanawake wa Tanzania (UWT), was of little help in assisting women off the job in their pursuit of income-generating activities. The women's union, which had grown out of the independence movement, was since its inception in 1962 financially dependent on the party and was organised along the same lines as the party. Although it had played an important role in galvanising women in the struggle for independence, it did not start from women's own associations as a basis for organisation, but rather superimposed its own notions of how women should be organised and what their concerns should be. While the UWT did attempt to address important needs in childcare, maternal health care, nutrition and income-generating projects, its top-down approach did not permit the popular mass base necessary in order to be truly effective and responsive to the needs of women.

The UWT's own organisation that was aimed at helping women with income-generating projects, SUWATA (Shirika la Uchumi wa Wanawake Katika Tanzania), concentrated its efforts in rural areas virtually to the exclusion of urban women. In fact, virtually no foreign donor or government assistance was available to urban women for income-generating activities. The UWT originally had its own department advising women on economic and financial activities, but for legal reasons decided in 1987 to replace it with SUWATA, which was registered as a company. SUWATA was charged with running projects to raise funds for the UWT; advising women on business projects and sources of credit; and serving as an intermediary between women's groups on the one hand and donor agencies and banks on the other. The views of SUWATA representatives reflected prevalent views within the UWT leadership:

> 'Women are ignorant of how to run small businesses and this is holding women back. They have problems keeping records and managing businesses. . . . Women do not want to trust each other and can't believe that they can do it. . . . The basic problem with women is lack of education. . . . They do not feel free and have no confidence to push ahead. They are not exposed to the world and meeting people and therefore lack confidence.'[18]

While there is undoubtedly some truth to many of the observations the SUWATA leader made, it would appear that her emphasis on the helplessness of women and neglect of the strength of women's own organisations is the kind of thinking that made it so difficult for SUWATA and the UWT to cultivate loyal supporters and provide meaningful organisation at the local level. One woman who owned a hair salon explained why she preferred working in independent women's organisations to the party-run organisation she belonged to: 'We work in the YWCA and our [income-generating] projects because we can use our creativity and do what we want.'[19] Another woman who had worked as a sewer for a YWCA clothing store for fifteen years said that eventually she would leave the store to work full time in her own tailoring business which she had started three years earlier, but that she would remain as a volunteer at the YWCA because, as she put it, 'I like the job a lot and feel attached to it and the people.'[20] These comments indicate how women preferred independent organisations, which they felt allowed for greater accountability, efficiency and individual input. Such organisations also served as barometers against which to measure the effectiveness of party and government organising initiatives and performance.

WOMEN'S ASSOCIATIONS IN TANZANIA

Tanzanian women have a rich history of association which has largely been unrecorded because most women's associations have tended to be small, informal and loosely organised. These are the kinds of organisations which rarely attract the attention of social scientists (except for anthropologists), nor do they gain the proper recognition they deserve. Women historically came together to organise feasts, funerals, weddings, and provide collective assistance during childbirth and illness. They formed savings clubs, beer clubs, loose associations of instructresses of young girls (for puberty rites), ritual cult groups, informal social and religious groups, mutual support groups for cultivation and harvesting, and age and neighbourhood groups.[21] John Iliffe noted in *A Modern History of Tanganyika*, for example, that most of the fifty-eight dance societies seeking registration in Dar es Salaam in 1954 were women's *lelemama* exorcism groups.[22] Women's economic associations in Tanzania, however, were never as large or varied as those found in West Africa, the powerful unions of traders, for instance. Instead, their involvement in micro-enterprises was characterised by co-operation with a complex array of individuals who ranged from relatives and friends to neighbours and people from the wider communities.

Although there are a few independent national women's organisations, like the YWCA, the Legion of Mary, or the new Association of Businesswomen of Tanzania, both the party and the government served as militating factors against the formation of broader associations. Following a pattern found in much of Africa, the newly independent state led by the single party gradually co-opted, absorbed and even eliminated independent organisations, especially economic organisations like the agricultural co-operatives and the trade unions, which they felt posed the greatest threat to state-led development. The government in Tanzania, like many other African governments, is gradually becoming more tolerant of the non-governmental sector and is in many cases encouraging its growth. Non-governmental associations became especially important on the African political scene in the 1980s and became one of the most important forces promoting greater pluralism and democracy and more bottom-up approaches to development in contrast to the heavy top-down government-dominated approaches of the past. Even the ruling party in Tanzania has acknowledged the need for independent organisations. In July 1990 former Party Chairman Julius Nyerere announced that the national trade union and the Co-operative Union were to become independent of the party. The national women's and youth organisations were to remain temporarily under the party's control but should also try to become independent. He addressed the General Council meeting of the women's union, saying that it should not be worried by the many women's groups springing up. Instead it should welcome them since they were also helping to serve the people.[23]

While the party and government may have inhibited the growth of larger, more diverse women's organisations in the past, they were unable to prevent the growth of smaller informal organisations. The circumscribed character of women's organisations in a country like Tanzania makes it all the more imperative not to simply look at participation in formal or visible political and economic institutions for it is in the realm of informal associations and networks that one begins to see more fully the scope of women's interactions.

The growth of women's associations in Tanzania is part of a broader urban response to economic crisis and to economic reform in which a wide variety of organisations are represented. The growth of economic organisations like rotating credit associations and independent co-operatives accompanied the flourishing of the informal economy that expanded, in part, because of sharp declines in real wages. Other groups formed to meet new demands that had arisen as a consequence of the changes in the economy, for example, parents' organisations seeking to meet new childcare needs because of women's increased involvement in self-employment. Business associations like the Tanganyika Chamber of Commerce, Dar es Salaam Chamber of Commerce, and the Metal Engineering Industries Development Association expanded their activities, and new organisations like the Association of Businesswomen were formed to create an environment more conducive to the private sector. Such organisations focused on problems ranging from trade liberalisation to the investment code, credit availability, devaluation, customs taxes and other such issues. Some organisations found themselves attempting to get the state to loosen up its restrictions on small informal enterprises and various sideline businesses. For instance, pressure came from the Medical Association of Tanzania to allow physicians to practise medicine legally and privately after putting in their hours at government hospitals. Others emerged to provide services and other resources the state could no longer deliver. For example, local defence teams took over where the police were corrupt or no longer effective. Finally, in this period, organisations which had previously been banned re-emerged since they could now take advantage of the few new political spaces which had opened up.

ROTATING SAVINGS SOCIETIES

One of the most pervasive forms of organisation that emerged with the burst of economic activity was *upato* or rotating savings societies, to pool and save money. These associations are found throughout the world, not only the Third World, but also the US, especially within newer immigrant communities, and in countries like Italy, England and Canada.[24]

Women historically participated in such societies in parts of Tanzania, but were more likely to pool clothes and food rather than money. These societies

visit each participant daily at her home or place of work if it was in the neighbourhood to make the collection, wrapping the money in the corner of the *khanga* (wrap-around cloth) that she wore or filling a small pail, basket or some other container with the money. This person, called *kijumbe* or secretary-treasurer, might claim a portion of the kitty for her service.

Women's involvement in *upato* societies not only reflects the increase in women's involvement in income-generating activities, it is also an indication of their greater control over household resources. One of the main purposes of saving in such a society was to get the money out of the house so as to keep the men of the household from laying their hands on it. Since women had come to bear the greater burden of paying for most daily household expenses, money earned from income-generating projects needed to be guarded carefully. Women were responsible for feeding the children and paying school fees and health care costs. Women were usually the ones to save and initiate the building of a house in those households in a position to do so. They also used their *upato* savings to reinvest in their projects. Occasionally if they had money to spare, they would treat themselves to new *khangas* or get their hair fixed in one of the new designs that had become popular with the increase in the number of hairdressing salons.

It is important to emphasise that while women were often perceived by both men and women as more capable of saving than men, the reason for this was the fact that it was primarily the woman's responsibility to make sure that the needs of the household and children were met.

WOMEN'S CO-OPERATIVES

Although most people owned and operated their own income-generating projects, this did not preclude becoming involved in larger income-generating associations and co-operatives. In Mlingotini, a small fishing village outside Dar es Salaam, women had been involved in transporting and selling coconuts to the city; growing and selling vegetables; making and selling pastries; and frying fish for sale. Twenty-one of these women formed a co-operative to mill rice, while continuing their own individual projects. Then in 1985 with the profits obtained from the mill, each woman contributed 150 Tsh ($1.50), and collecting around 3,000 Tsh ($30) they started a business transporting and selling fish to the city.

The leader of the co-operative said she had witnessed several changes over the years that preceded the formation of the co-operative. The women in her village were engaged in small businesses like never before and they were for the first time keeping all their own profits. She said: 'We women are getting clever. We didn't do that in the past.' She added that men would have liked to object to their wives' increased economic activity, 'but they keep quiet. They know they can't support the family and they need the woman's income. Some men are getting ambitious themselves and are trying to find businesses for themselves.'[26]

The women jointly purchased nets and a *hori* (dugout boat). 'When the women got nets they didn't want them transported by car back to the village. They wanted to walk them home on their heads so everyone could see. They were so proud,' she said.[27] At first they bought fish caught by a trawler owned by the nearby Norwegian-Tanzanian training centre, Mbegani Fisheries Development Centre. Later when the trawler ran into difficulties and was no longer operating, they hired a fisherman and sold the fish he caught.

In villages on the outskirts of Dar es Salaam women were involved in similar collective endeavours: in Chiwanga women had organised to cultivate pineapples and later started a flour mill; in Mbegani women used money from a sewing project to start another flour mill; and in Mbweni, Pande and Mlingotini women had started their own co-operative shops, and a fish smoking co-operative in Mlingotini.[28] Examples like this reflect the varied and collective strategies women pursued in their efforts to make ends meet.

It should, however, be pointed out that co-operatives were not always women's first choice in organisational arrangements for their projects. Most women, including the ones mentioned above, carried out their businesses alone or with one other person. Participating in a co-operative would have been only a secondary venture. Rarely would women have relied on a co-operative for their primary source of subsistence. Even better-off women formed co-operatives mainly as a way of gaining access to credit or loans since foreign donors as well as government agencies would approve funds only for co-operatives.

ASSOCIATION OF BUSINESSWOMEN IN TANZANIA

The creation of the Association of Businesswomen in Tanzania in January 1990 is another indication of the changing economic role of women in Tanzania. It was not only the poorer women who had begun to engage in income-generating businesses in the 1980s; professional and middle-income women were also leaving their salaried positions to go into business or were involved in sideline enterprises. They had established large tailoring businesses, dry cleaning companies, flour mills, secretarial service companies, hair salons, export and import businesses, bakeries and other small manufacturing and service industries. The pervasiveness of such relatively lucrative businesses among women was unprecedented, representing a phenomenon which began to emerge in a significant way only in the mid-1980s.

It is interesting that the businesswomen's association, although initiated by middle-class and big businesswomen, included as part of its constituency poor women involved in income-generating activities and wanted to form with them a 'strong business community' of women.[29] The organisation was seeking to gain recognition from the country's financial institutions, which had neglected women and their credit needs. The organisation also planned to provide free expertise and conduct seminars for women on issues relating to marketing and other business skills. It also began to investigate possibilities of acting as an intermediary agency to obtain loans for local women's organisations that find it difficult to make such applications themselves.

Not long after its inception, the organisation found that economic possibilities for women were severely constrained by the weak representation of women in the political arena. Leaders of the association became vocal in criticising the official party-affiliated women's union, arguing that this union was not strong enough to represent the new interests of women in Tanzania. They began to express the need for stronger women's participation in parliament. The experience of the businesswomen's association suggests that the strengthened economic clout of better-educated middle-class women will eventually have important repercussions in the political arena, leading to greater political mobilisation of women in the country.

PARENTS' ASSOCIATIONS

Women had also been active in organisations to form day-care centres. The need for such organisations emerged as a result of women's increased involvement in small businesses and their demands for new childcare. The lack of day-care centres had been one of the main obstacles in keeping women out of employment, but creating such day-care centres had never been a priority of the government.[30] The Ministry of Labour and Social Welfare, along with the women's union UWT, was responsible for creating and supervising all day-care centres after 1976, but their efforts were minimal.[31] By 1980 there were only 2,996 day-care centres in the entire country with 94 in Dar es Salaam.[32] As more women of all income groups became self-employed, they became increasingly involved in initiating parents' organisations to form nurseries.

One city council administrator, who also had a sideline project raising chickens, formed a parents' organisation in her neighbourhood of Kijitonyama in Dar es Salaam. Together they built the school itself, each paying for the construction of a window, door or other part of the building. They also jointly paid for the furniture, equipment and the teacher's salary.[33]

In other instances, women's religious organisations sponsored the creation of day-care centres. The Young Women's Christian Association had been especially active in this area. In Uzuri, Manzese, for example, they had started a nursery for 70–80 children run by two teachers. The parents paid 30 Tsh ($0.30) a month and provided uniforms.

In Mferejini, Manzese, women were active in an organisation that was formed to start a nursery school, which received no help from outside. They used the premises of the party branch office which had only two small rooms. Parents paid a 20 Tsh ($0.20) fee per child each month and provided pencils, notebooks and other supplies. Eighty children crowded each day into two small rooms. Apart from a blackboard and one desk and chair, there was no other equipment, furniture or toys in the school. Nevertheless, this day-care centre was one of the most memorable places I visited during my fieldwork in Manzese. What made it so special was the 16-year-old woman who had put all her energy into teaching the children everything she had learned in her seven years of primary school. I was astounded to see 4- and 5-year-olds sitting on the floor working out complicated multiplication and division problems. Even the smallest children could read and write.

CREATING NEW NETWORKS

In addition to women's associations, women built networks around their businesses with people they trusted for information, advice and assistance of various kinds. Women generally owned their own projects (94 per cent). Forty per cent worked alone in their projects, 61 per cent were assisted by their children, while 50 per cent worked with one or two others in addition to their children. For the most part, when they did collaborate with someone else in their small business, they worked with another family member, distant relative or friend who perhaps had a project of her own.

However, women increasingly found themselves having to go outside their closest family and kin to broaden their circle of contacts in order to do business. Neighbours in the city, usually of different ethnic origin, were fast becoming a part of this new circle of close associates. Often the same mutual obligations and expectations reserved for kin were extended to neighbours and friends, and the trust engendered by such relations became crucial when these same people were involved in a joint project. As Christine Obbo writes of urban women in Uganda:

> Fictive kinship usually developed among people who co-operated intimately in day-to-day activities, paralleling the moral, economic and social support relatives give each other in the rural areas. As one Ganda woman said 'your neighbours are your relatives'. This use of pseudo-kinship was also recorded among other groups. . . . Any group of co-workers or neighbours could address each other in kinship terms.[34]

I realised this from the outset when carrying out my interviews. Expressing great affection for their female friends and neighbours, women would often insist on being interviewed with them. Sometimes this involved going to great lengths to make arrangements so that all could be present for the interview. Relations with neighbours and friends, which in the past were reserved for helping each other with household tasks, tending children and plaiting each others' hair, were increasingly being extended to the area of micro-entreprises.

Urban life presented new opportunities for women to meet new people and establish information and resource networks. For example, living together in one house, as many families did, encouraged women neighbours to start projects together. It was not uncommon to find women living in the same house sharing a project like making local beer, while maintaining separate accounts. In one apartment complex, I visited four separate apartments where women were involved in sewing as a business and I was told that there were more women sewers in the building. They were all friends and had inspired each other to begin their projects to supplement their husbands' meagre wages. They helped each other out with their businesses and insisted that they did not compete with one another. Other women found their formal jobs a place to exchange ideas and support for their projects. The hairdressing salon was also another frequently mentioned place for meeting other women to discuss business.

Working arrangements with friends and neighbours were of various kinds and served many different purposes. Friends would accompany each other to a restaurant or bar, one perhaps selling fried fish and chips and the other selling *makongoro* soup. While companionship was obviously an important part of such an arrangement, it also enabled them to help each other out in the more practical aspects of buying inputs and preparing the food for sale. Some reported they felt safer with a friend if they had to sell late into the night.

Women market sellers tended to sell in the same part of the market so they could help each other out with childcare if necessary and tend each others' stalls if one had to go away temporarily. They could also form a warning and support network to help each other evade the city council militia that came around to check licences. Since most women sold their wares without a licence, this could be a potential problem. Many women, for example, sold maize and flour in rows outside the market because they could not afford the licence and fees required to obtain a market stall. They felt the unpredictability of their business did not warrant procuring a licence. Selling together gave them a sense of security and mutual support in case they were caught.

CONCLUSIONS

The decline of large-scale industry in Tanzania throughout the second half of the 1970s and the 1980s resulted in declining real wages and layoffs, putting greater pressures on household members to pursue income-generating projects. Similar patterns were found in many African countries to one degree or another. Urban women thus moved from a position of relatively little involvement in income-generating activities, to being, in many cases, the main economic support for the household. Women who were employed often had to leave their jobs because the pay was too low or they had to seek sideline incomes. Women's expanded economic role led to an increase in the variety of women's economic organisations and associations that could cater to women's new demands as a result of their businesses.

The forms of organisation women adopted were indicative of their needs and priorities. These priorities were not always reflected in party-affiliated structures that offered superimposed blueprints of organisational structures women were to adopt. While the party and government had limited larger and more formal independent associations in order to ensure government control of the economy, women found little difficulty in organising their own local informal associations. By the late 1980s it had become evident that the state's posture towards the non-governmental sector was untenable and its inability to restrict independent organisational activity coupled with a realisation of the need to bring a broader array of forces into the development process gave way to a greater openness to organisations like the newly emerging Association of Businesswomen in Tanzania.

In order to understand women's response to the impact of deindustrialisation upon relationships in Tanzanian society it is important to look not only at such visible formal organisations, but also at those organisations embedded in the daily lives of women. Through such associations, women are transforming their lives in their day-to-day struggle to survive and at the same time changing the political landscape of the country.

NOTES

1 See for example Victoria Daines and David Seddon, 'International Development Policy and the Recession', in *Survival Struggles, Protest and Resistance: Women's Responses to 'Austerity Programmes' and 'Structural Adjustment'*, forthcoming; Jane Jacquette (ed.) *The Women's Movement in Latin America: Feminism and the Transition to Democracy*, Unwin Hyman, Boston, Mass., 1989.
2 Michael Lofchie, 'Tanzania's Agricultural Decline', in Naomi Chazan and Timothy Shaw (eds) *Coping with Africa's Food Crisis*, Lynne Rienner, Boulder, Colo., 1988, p. 144.
3 United Republic of Tanzania, *Economic Survey 1980–81*, Government Printer, Dar es Salaam, 1989.
4 N.H.I. Lipumba and Benno J. Ndulu, 'Long Term Trends in Exports', *Tanzanian Economic Trends, 1989*, 2(1), pp. 11–12; A. Mbelle, 'Industrial (Manufacturing) Performance in the ERP Context in Tanzania', *Tanzania Economic Trends, 1989*, 2(1), pp. 7–9.
5 Mbelle, 'Industrial (Manufacturing) Performance in the ERP Context in Tanzania', p. 7; Samuel Wangwe, 'Recovery of the Industrial Sector: Some Lessons from Experience', *Tanzania Economic Trends*, 1988, 1(3), pp. 34, 41.
6 Jannik Boesen, Kjell J. Havnevik, Juhani Koponen and Rie Odgaard (eds) *Tanzania: Crisis and Struggle for Survival*, Scandinavian Institute of African Studies, Uppsala, 1986, p. 19.
7 Mbelle, 'Industrial (Manufacturing) Performance in the ERP Context in Tanzania', pp. 9–10.
8 Bureau of Statistics, *1988 Population Census: Preliminary Report*, Ministry of Finance, Planning and Economic Affairs, Dar es Salaam, 1989.
9 Bureau of Statistics, *Household Budget Survey: Income and Consumption 1976–77,* Ministry of Finance, Planning and Economic Affairs, Dar es Salaam, 1977.
10 International Labour Organisation, *Basic Needs in Danger. A Basic Needs Oriented Development Strategy for Tanzania*, International Labour Organisation, Addis Ababa, 1982, p. 143.
11 Anna K. Tibaijuka, *The Impact of Structural Adjustment Programmes on Women: The Case of Tanzania's Economic Recovery Programme*, Economic Research Bureau, University of Dar es Salaam, Report prepared for the Canadian International Development Agency (CIDA), 1988, pp. 31–2.
12 Aili Mari Tripp, interview 15 August 1987.
13 R.H. Sabot, *Economic Development and Urban Migration: Tanzania 1900–1971*, Clarendon Press, Oxford, 1979, p. 92.
14 Tibaijuka, *The Impact of Structural Adjustment Programmes on Women: The Case of Tanzania's Economic Recovery Programme*, p. 37.

15 Aili Mari Tripp, interview 11 October 1987.
16 Aili Mari Tripp, interview 18 August 1987.
17 See M.H. Mgaya, 'A Study of Workers in a Factory', MA thesis, University of Dar es Salaam, 1976.
18 Aili Mari Tripp, interview 28 September 1987.
19 Marja-Liisa Swantz, interview 23 May 1989.
20 Aili Mari Tripp, interview 17 August 1987.
21 Marja-Liisa Swantz, *Women in Development: A Creative Role Denied?*, C. Hurst and Company, London, 1985, p. 160.
22 John Iliffe, *A Modern History of Tanganyika*, Cambridge University Press, African Studies Series, Cambridge, 1979, pp. 391–2.
23 *Daily News*, 17 July 1990.
24 See for example, S. Ardener, 'The Comparative Study of Rotating Credit Associations', *Journal of the Royal Anthropological Institute*, 1964, p. 94; Clifford Geertz, 'The Rotating Credit Association: A "Middle Rung" in Development', *Economic Development and Cultural Change*, 10(3), 1962; D.V. Kurtz, 'The Rotating Credit Association: An Adaptation to Poverty', *Human Organisation*, 32(1), 1973; C. Velez-Ibanez, *Bonds of Mutual Trust: The Cultural Systems of Rotating Credit Associations among Urban Mexicans and Chicanos*, Rutgers University Press, New Brunswick, 1983.
25 Marjorie Mbilinyi, 'A Review of Women in Development Issues in Tanzania', Report for World Bank, mimeograph, 1990, pp.72–6; Anna K. Tibaijuka, Mary Kisanga, Joyce Hamisi and M. Abubakar, 'Strategies to Enhance Women's Access and Effective Utilisation of Institutional Credit: Report on a Survey on Women's Land and Property Rights in Tanzania', Presidental Commission of Enquiry into Monetary and Banking Systems in Tanzania: Sub-Committee on Women's Credit Facilities, 1979, mimeograph.
26 Aili Mari Tripp, interview 4 June 1988.
27 Aili Mari Tripp, interview 4 June 1988.
28 Marja-Liisa Swantz, *The Role of Women in Tanzanian Fishing Societies: A Study of the Socioeconomic Context and the Situation of Women in Three Coastal Fishing Villages in Tanzania*, Institute of Development Studies Women's Study Group, University of Dar es Salaam, 1986, p. 57; Marja-Liisa Swantz, interview 23 April 1990.
29 *Daily News*, 19 March 1990.
30 Swantz, *Women in Development: A Creative Role Denied?*, pp. 150–1.
31 Susan G. Rogers, 'Efforts Toward Women's Development in Tanzania: Gender Rhetoric vs. Gender Realities', in Kathleen Staudt and Jane Jacquette (eds) *Women in Developing Countries: A Policy Focus*, Haworth Press, New York, 1982, pp. 28–9.
32 International Labour Organisation, *Basic Needs in Danger. A Basic Needs Oriented Development Strategy for Tanzania*, p. 145.
33 Marja-Liisa Swantz, interview 23 May 1989.
34 Christine Obbo, *African Women: Their Struggle for Economic Independence*, Zed Press, London, 1980, p. 115.

Chapter 7

Strategies against sweated work in Britain, 1820–1920

Sheila Rowbotham

ABSTRACT

In the nineteenth century terms like 'slop-work' and 'sweating' were used to describe a system of subcontracting and low-paid, unorganised forms of employment characteristic of industries like clothing where a pool of cheap labour combined with high urban rents and markets which required flexibility in production methods. The sweating system developed both out of decayed handicrafts and alongside factory production. It was labour-intensive work in which there was little incentive to mechanise.

Sheila Rowbotham draws on material from Britain to show the range of strategies adopted in opposing sweating in the period 1820–1920. By the early twentieth century the campaigners against sweating, who saw it as not only exploitative but retrogressive, were able to secure laws to fix rates in several low-paid industries. The case for abolishing homework was often linked to welfare policies concerned to improve the health of mothers. Though some reformers combined state measures with attempts to organise sweated women workers, the emphasis came to be placed upon regulation by the state.

INTRODUCTION

The 1980s growth of casualised employment in the British economy coincided with a shift in prevailing economic theories and policies. Broadly, from the late nineteenth century, working conditions had become increasingly regulated by law and by recognised trade unionism. An influential lobby among employers had accepted a regulated economy on the grounds that official trade unionism and negotiated settlements were less disruptive than unofficial militancy. During the 1980s, this was to be challenged and the advantages of a 'flexible' workforce stressed. Right-wing arguments for flexibility have stressed the need to cut labour costs in order to be competitive. But 'flexibility' also found left-wing advocates who saw a shift from large-scale factories on the Fordist model to a post-Fordist organisation of production as creating liberatory 'new times' for workers released from the discipline of Taylorism. They tended to slide from a description of economic trends into an uncritical celebration which did not examine the actual implications of flexibility for casualised workers. On the other hand, some critics of post-Fordism have simply condemned the new conditions of employment instead of acknowledging that they have any positive features or asking what kind of organisation might be developed in changing circumstances.[1]

The erosion of any consensus on the proper relationship between the state and production makes it possible to look at changes in the organisation of work and the emergence of state intervention in the economy in the nineteenth and early twentieth centuries in a new light. This alters some of the assumed perspectives in the current debate. A historical examination of women's casualised work questions the assumption, for instance, that 'flexibility' is a modern invention; the supposedly 'new' forms of production actually have very old precedents. Similarly the inequitable consequences of market forces are evident. The efforts of late nineteenth-century reformers to regulate the economy in order to protect vulnerable groups of workers provide plenty of evidence to correct a facile optimism about the market as a guarantee for creative autonomy. On the other hand the particular combination of state intervention from above and a mixed economy can be seen to arise out of specific social circumstances. It is not the only conceivable alternative to the market. A historical approach moreover also shows that ideas about who can be organised have changed over time. It is erroneous consequently to impose universal assumptions about 'correct' forms of labour organising, which are in fact based on a specific and limited period of industrialisation without examining differing working situations. There have been many differing forms of organisation over two centuries. This is particularly true in relation to women workers who have not always corresponded to the male model of employment.

Recent studies on gender and labour have revealed that while the woman worker in clogs and shawl going to the factory in the early morning is popularly associated with the 'industrial revolution' in Britain in the early nineteenth century, in fact most working women were to be found in outwork or in small workshops, employed as domestic servants or providing services like laundry work. By the middle of the nineteenth century women's opportunities for employment in Britain had actually narrowed from a hundred years before. Contemporaries were also concerned about a new kind of casualised work, or 'slop-work', which was particularly prevalent in London where rents were high and factories did not get established. Work processes could be broken down and done by cheap labour at home or in small workshops, and subcontractors farmed out work.[2]

Towards the end of the nineteenth century this system of production came to be called sweated work. The term 'sweating' had a range of meanings. It was used to describe work which was subcontracted either to workshops or individuals. More generally it described employers who paid low wages and whose workers had bad conditions. Payment by piece-rate also meant workers were struggling against time and always under pressure, often still only just earning a livelihood. Sweating was frequently seen as caused by the system of subcontracting and the sweater or middleman was blamed for the low rates. But systematic investigations by Beatrice Potter (later Webb) and Charles Booth in the 1880s and 1890s showed that they were being made scapegoats in generally oppressive working conditions; also not all homeworkers were regarded as sweated; there were still domestic industries in the countryside based on outwork. Moreover, some homeworkers were the wives of skilled workers or clerks and did not fit into the downtrodden stereotype. Nor was homework a completely separate form of production. There was often a connection between homework and factory work. Some homeworkers were former factory workers. Some factory workers took work home and women could do a variety of jobs. Sweated homework, then, was only one aspect of a wider spectrum of casualisation.

Contemporaries blamed slop-work and sweating on over-population, immigration, women working and greed. Beatrice Potter and Charles Booth were aware, however, that it arose with a large pool of cheap labour and high rents, when it was possible to break down the stages of production so they required little skill. Booth also saw it was linked to specific markets.

> In the present stage of meteorological knowledge the demand for umbrellas and sunshades cannot be calculated to a nicety beforehand; the time at which men will order ties and the particular reasons they have for preferring one kind to another are facts shrouded in mystery.[3]

The flexibility of the small undercapitalised unit of production suited the employer producing for the fashion market. By the end of the nineteenth century new ways of retailing clothes also gave an incentive to produce cheap goods for a larger market. Sweating could arise out of the decline of handicrafts or in connection to the work generated by factory production.

As Marx noted,

> Beside the factory operatives, the manufacturing workmen and the handicrafts men, whom it concentrates in large masses at one spot, and directly commands, capital also sets in motion, by means of invisible threads, another army; that of the workers in domestic industries, who dwell in the large towns and are also scattered all over the face of the country.[4]

Beatrice Potter settled for a generalised definition of sweating in 1881: 'all labour employed in manufacture which has escaped the regulation of the Factory Acts and the trade unions.'[5]

In the sweated trades there were, along with clothing workers, a great host of artificial flower makers, box makers, brush makers, book folders, paper bag makers, wood choppers, envelope makers, cigar and cigarette makers and wrappers, ostrich feather curlers, lace makers, straw-plait makers and many more. Though a service industry, laundry work was also described as sweated work.

Beatrice Potter was to be an influential voice in a wider process from the 1880s onwards in which such categories of workers were being defined as unorganisable. By the early twentieth century reformers, trade unionists and modern-minded big employers were arguing that sweated work should be abolished. They tended to see it as a backward-looking form of production which prevented development and modernisation, rather than a system of production which drew on particular labour markets in certain economic contexts.

New questions of the past are stimulated by the present. The success of SEWA (see Chapter 5) enables us to reconsider the range of strategies adopted in the nineteenth century before the conviction formed that state regulation was the *only* realistic way of tackling the oppressive aspects of homework. It raises the question of why the awareness of the need to *combine* organising with pressure on the state recedes in relation to the sweated homeworkers from World War One and remained forgotten for so long. This is a big question to answer. But even by posing it the history of submerged casualised workers appears in new relief. An even bigger question lurks behind the attempts to mobilise and gain access to resources controlled by the state. What changes in the power relations within society and state would transform the conditions of casualised workers?

In western capitalism organised workers came to accept the rules of the game and bargained within the terms of the factory system. Although a shift towards 'flexibility' and 'casualisation' during the 1980s have tended to lead to a worsening in conditions on the whole and have appeared to block labour's tried strategies, they also have created new circumstances which present possibilities. A reorganisation of production and not simply the *defence* of workers has been raised again. In whose interests and on whose terms economic restructuring and development should be conceived and put into effect is the crucial issue.

Particular experiences cannot be transposed mechanically from one context to another. The nineteenth- and early twentieth-century attempts to organise around homework did not confront a modern globalised economy, for example. However, while the term 'sweating' is not synonymous with 'casualisation' in the modern sense, there are sufficient overlaps to make for comparison. Although state regulation came to be seen as the best way to deal with the problems of sweated work between the 1920s and 1970s, there is in fact an earlier history of organisation in which a wide range of strategies were tried: co-operatives, trade unions, philanthropic schemes. Some aimed to defend workers, others to create quite different forms of production.

STRATEGIES

Co-operative association

Co-operative associations in the first half of the nineteenth century were, like trade unions, a means by which workers sought to improve conditions and develop the habits of democratic organisation. Thinkers like William Thompson and Robert Owen envisaged large-scale co-operative enterprises and continually developing technologies as an alternative to capitalist development. Working-class co-operators attempted zealously to put this alternative vision of economic development into practice by setting up communities, not only in Britain but in France and the USA.

By the second half of the century supporters of co-operation were settling for smaller-scale measures. Co-operatives were being presented as an intervention within capitalism which could alleviate poverty and put an end to particular evils like slop-work. This form of co-operation was really philanthropy with a self-help approach. The Christian Socialist J.M. Ludlow returned from Paris in the late 1840s enthused by French workers' co-operative associations as a means of combating the slop-trade.[6] Henry Mayhew's articles in the *Morning Chronicle* exposing conditions also made a deep impression on the Christian Socialists, inspiring Charles Kingsley's tract, 'Cheap Clothes and Nasty', and his novel about working-class conditions and politics, *Alton Locke*.[7] In 1850 a Christian Socialist Brotherhood was set up and embarked on a series of co-operative ventures, including a Needlewoman's Association for the relief of slop-women. However, this was not a self-governing co-operative like those formed among male workers by the Christian Socialists or indeed the French women's associations which in 1848 had bargained for homeworkers. A superintendent gave out work and had the power of dismissal, 'subject to the ladies' committee or the lady visitor of the day'.[8]

There were various subsequent attempts to form philanthropic co-operatives for needlewomen. In 1860 an Institution for the Employment of Needlewomen was founded. By 1861 500 women were employed in light airy rooms. However, the only groups reached were wives of shopkeepers, decayed gentlewomen and ex-governesses, not the poor slop-women. This project did improve the conditions of the needlewomen employed but foundered because it had not attended to demand.[9]

A series of trade union-backed co-ops failed in the 1870s and the working-class co-operative movement shifted towards distribution but there was a revival of interest in producer co-ops from the 1880s among middle-class reformers. Profit sharing and co-partnership were seen as alternatives to trade union militancy.

A venture in east London to provide a co-operative alternative to sweating shows such ventures could be risky. Distressed by stories of sweated workers, a merchant W.J. Walker decided to act as a middleman for the needlewomen without taking any profit at all. He discovered a machinist, if competent, could make 10 to 15 shillings a week but that a finisher still could not make a living.[10] In 1888 he saw a letter in *The Times* from Francis Peek about the East End branch of the Labour Aid Association arguing that increased pay for needlewomen would result in 'the cheaper labour of Germany, France and Spain swamping our market with ready-made goods'.[11] It called attention, however, to the outlays made by needlewomen who had to pay an exorbitant price for small amounts of thread and buttons and the hire of sewing machines. The Labour Aid Society had decided to start a limited company which was to develop into a co-operative society in which workers would have a share. Walker and Peek provided the capital for a light airy factory without charging interest. One half did tailoring, the other made shirts. After a year the tailoring side paid but the shirt making did not. Walker concluded that the only remedy was the abolition of all homework, despite the suffering this would cause to women who could not do other kinds of work. He also advocated trade unionism. Similarly the Select Committee on the Sweating System in 1890 was to recommend 'the extension of co-operative societies, and . . . well-considered combination amongst the workers'.[12]

A dispute broke out in the socialist movement in the early 1890s about co-operation as a solution for casualised women workers. In 1890 an attempt was made to form a Laundry Women's Co-operative Association. It was announced in the newspaper of the Socialist League, *Commonweal*, that they wanted socialists to buy shares or send them laundry. The aim was to stop sweating in the laundry trade. 'We also make a special appeal to our comrades as women, for not only do women suffer as wage slaves but as chattel-slaves also.'[13]

Amie Marsh, Jenny Willis, Ida McKenzie and Edith Lupton were involved in the co-operative. However, the *Commonweal*'s anarchist-communist politics were too absolute to concede that the co-op was worth bothering about and the editors disapproved of 50 per cent interest being paid on the shares. One correspondent, probably Edith Lupton, a middle-class woman with a background in radical local government, insisted that interest or no interest the co-op was 'honestly' trying to alleviate the 'terrible trade conditions under which laundry women are at present suffering'.[14] *Commonweal* replied from a great height.

> We are not sure that co-operation, or even trade-unionism will sweep their slavery away. They may both do something to improve the conditions of some of these workers, but nothing but the Social Revolution will raise the mass from the horrible misery from which most working women suffer at the present.[15]

Meanwhile the Women's Trade Union League led by the Liberal Lady Dilke had got involved in the laundresses' agitation. In April 1891 they reported a meeting to consider how to extend protective legislation to laundries.[16] By July the laundresses had got support from twenty-seven trades councils. Several thousand laundresses had attended the first working women's demonstration in Hyde Park along with railway workers. Tom Mann, one of the leaders of militant new unionism, had addressed them along with May Abraham, an organiser from the Women's Trade Union League.[17]

Eleanor Marx writing in the Austrian socialist paper *Working Women's Journal* in May 1892 remarked that the laundresses had sent a delegation to Parliament to demand coverage under the Factory Acts. She added, with sweeping contempt,

> It is worthwhile to make the point that immediately Mrs Fawcett the reactionary bourgeois advocate of women's rights (of the rights of property-owning women), who has never worked a day in her life, along with Miss Lupton, an anarchist (likewise a woman of the middle class), sent a counter-delegation to protest against this intervention in woman's labour.[18]

However, in contrast Eleanor Marx, herself of course a woman of the middle class, praised the Women's Trade Union League organiser, the middle-class May Abraham: 'It is largely thanks to her that these women now clearly understand the urgent question of governmental limitation of the work day'.[19]

There were some pretty heavy guns firing over the heads of the laundresses. Their cause revealed deep disagreements among socialists, anarchists and feminists about how workers could bring about changes. Behind the infighting, strategic issues were at stake. Eleanor Marx was fiercely opposed to the anarchist-communists who had taken over the Socialist League and *Commonweal*; and she had little time either for William Morris, the founder of the league, who was isolated and bewildered, opposed both to using Parliament and to the intransigence of the anarchist-communists. The anarchists too had their disputes; while some insisted that the only aim was an all-out insurrection, others supported co-operatives and communes. Among liberal feminists there was a tension between Millicent Fawcett's anti-state views and opposition to protective legislation and Lady Dilke's approval of campaigning for legislation. Eleanor Marx's criticisms of Mrs Fawcett and Miss Lupton in the Austrian socialist women's paper have their context in the conflict between socialist women in Germany and Austria and liberal feminists over support for protective legislation as a means of improving women's working conditions.

In retrospect, the early 1890s can be seen as a time when the pivots of assumptions were changing. There was ceasing to be a place for Morris's libertarian socialism; various forms of what he called state socialism were coming into ascendancy.[20] There was a marxist form with varying degrees of emphasis upon organisation from below and there was a fabian version, which while highly critical of the marxists' emphasis on class conflict, believed in a strong centralised state. Anti-statist strands became polarised among anarchists and liberals.

This theoretical shift has a bearing on how to end sweating. Beatrice Potter's *The Co-operative Movement in Great Britain* was published in 1893 and articulated an approach which the fabians were to push tenaciously for the next two decades. She dismissed voluntary association, either co-operatives or trade unions, as a viable strategy for the shifting poor.

> The hand-to-mouth existence of the casual labourer, the physical inertia of the sweater's victim, the vagrant habits and irregular desires of the street hawkers, and of the mongrel inhabitants of the common lodging-house – do not permit the development, in the individual or the class, of the qualities of democratisation and democratic self-government.[21]

Organisation was linked to trades where there was already state regulation: 'The wage earners of Birmingham and London at work in their homes, or in the workshops that escape regulation, are apparently incapable of association as consumers or producers.'[22]

A clear line is drawn between a respectable labour movement capable of negotiation and within the terms of the state's protection and casualised sweated workers. Although Beatrice Potter prided herself on a hardheaded approach and was speaking from her investigative experience, there is also a political point being made. The Webbs were to foster a particular concept of what constituted the labour movement and push for a political combination of labour and the state. As Stephen Yeo points out, this involves a presupposition about the relationship of society and the state: 'The State is not seen as a set of relations of production. It is only seen as an IT – an unproblematically good thing – to be captured, as its own rightful prize against the inefficient.'[23]

This uncritical viewpoint on the state was to have a profound effect upon the British labour movement and sections of the Liberal Party searching for a broader social base. Despite political differences it appeared to converge with the militant new unionists' demands for an eight-hour day.

It did not, however, go completely unopposed. Apart from the anarchists there was in the milieu of the Independent Labour Party, the feminists and the syndicalist left varying degrees of suspicion of state socialism. From these currents came rather different approaches to the predicament of sweated casualised workers and continuing attempts to organise co-ops and unions among these 'outcasts'.

The memory of the earlier radical vision of co-operation as an alternative social system was not entirely obliterated. In 1906 a Leeds socialist feminist with considerable experience of organising women clothing workers, Isabella Ford, quoted enthusiastically from a letter written in 1851 by French socialist women who had been active in the French co-ops during the revolution of 1848, Jeanne Deroin, and the schoolteacher arrested with her, Pauline Roland. They insisted, 'only by the power of association based on solidarity – by the union of the working classes of both sexes to organise labour – can be acquired . . . the civil and political equality of women and the social right for all'.[24]

Isabella Ford, a member of the Independent Labour Party, was not opposed to using the powers of the state but stressed the integral connection between mobilisation and pressure from below. However, by the 1900s the anarchist-communists' anti-statism had appeared in a new form, anarcho-syndicalism. The syndicalists were hostile to the state's intervention which Liberal and Labour supporters saw as progressive. Perhaps the Co-op advert in *The Syndicalist* in 1913 was a little opportunist: 'By dealing at our Co-operative Stores you strike at the great evil of sweating.'[25]

Consumer power was not the first priority for syndicalists who emphasised direct action at the workplace. The problem for casualised women was that male workers, for instance railway men and builders and engineers, were the focus of syndicalist militancy. Syndicalist politics, however, had a broader influence with important implications for women. For example the socialist feminist Sylvia Pankhurst became convinced that direct action had to be linked to the community as well as the workplace. She devised a whole series of approaches in the attempt to mobilise precisely the constituency Beatrice Webb had so firmly rejected – the brush makers and barmaids married to dock labourers of east London. Sylvia Pankhurst *combined* the use of direct action with pressure on the state. The campaign for women's suffrage had extended during World War One into campaigns around pay, rent, soldiers' allowances, prices. She also believed in setting up projects as alternatives with a practical side. Among these prefigurative models was a producers' co-operative making toys. However, there was a conflict between its profits and social purpose. It never paid for itself and had to be subsidised by wealthy supporters. Nor was it a harmonious co-op. The woman who managed it tried to impose business principles which went against the socialist feminist principles Sylvia Pankhurst endorsed.[26]

Nonetheless casualised women workers were mobilised in campaigning for the vote and more control over their work and daily life. It was not only in east London that poor women in this period showed themselves capable of dramatic action. High prices in Italy, Spain, Russia and France brought them into the streets in periodic mass strikes. Beatrice Webb's characterisation of them as simply victims and vagrants on the margins of modern society did not go uncontested.

Trade unionism

Unions, like co-operatives, have taken many forms. Trade unionism in Britain in the 1830s and 1840s was socially and politically ambitious, imagining that unions of the working class would transform capitalism. The groups who mobilised were not only factory workers combining against capital. For instance in 1840 a strike broke out among lace-runners or embroiderers. Four hundred and forty Nottingham homeworkers struck, not against the manufacturers, but against the middlewomen, their agents or 'mistresses'. A chain of up to three of these agents existed between the homeworkers and the manufacturers in the lace trade. During the recession the lace-runners sought to limit the amount the agents took to a penny in the shilling from their earnings. They recognised their role was necessary and had a concept of a just reward 'for the trouble they take in going with our work to the warehouses'.[27] They asked the manufacturers to price the lace-pieces and some began doing so. However, the agents refused to yield and the lace-runners did not win though they managed to maintain their association.

These early women's unions had social as well as economic aims, an approach which reappeared in the 1870s when Emma Paterson established several small unions among women in skilled trades like bookbinding. Emma Paterson was of middle-class origin but had had to earn her living as a bookbinder and was married to a skilled worker, a cabinet-maker. She was inspired by women's benefit and trade union societies in the United States, particularly umbrella makers and typographers.[28] She also admired the Christian Socialists' work in campaigning against the conditions of needlewomen and slop-women, and invited the elderly Charles Kingsley to attend the first conference of the Women's Protective and Provident League (later Women's Trade Union League) in 1874.[29]

Emma Paterson believed in a cautious, friendly society approach as a means of raising women's pay and conditions. She stressed the self-emancipation involved in organisation and its democratic potential, and was dubious of the benefits of state intervention through protective legislation. The league had various offshoots, a Halfpenny Savings Bank, the Women's Labour Exchange, a workers' restaurant and a swimming club. It was provident as well as protective.

In the 1870s outworkers were an accepted part of the trade union movement. At the TUC conference in 1877 Emma Paterson found an ally in her opposition to proposals to extend the Factories and Workshop Act to cover homework, from the Leicestershire Seamers' and Stitchers' Society, a union of homeworkers which the league had helped to form. Their representative, Mrs Mason, told congress: 'There were a great many villages connected with their Union, and the work they undertook was never done but in their own houses'.[30] Some women 'worked as a pastime', others like herself because their husbands were sick and they supported the family. Increased wages would reduce the hours they had to work.

State intervention would mean inspectors popping into people's homes but 'there were times when it was hardly desirable to have their houses inspected'.[31] But she brought the house down by adding that she did favour a law which would provide inspectors to make sure lazy husbands were out of bed before lunchtime!

In her account of *Women and Trade Unions*, Sheila Lewenhak points out: 'In villages with a stationary population homeworkers not only could organise, they had organised all through the first century and a half of the development of trade unionism'.[32] It was a time-consuming, exhausting effort trudging across the countryside after work in all weathers and sadly Mrs Mason died young. However, it was not true that scattered workers were unorganisable.

In the 1880s and 1890s the league continued its work. In 1885 Emma Paterson visited her friend, a middle-class Quaker and feminist in Leeds, Hannah Ford, where she recruited her daughter Isabella to help with a Tailoresses' Society formed by the league.[33] The all-women societies were supported by Liberal trade union men, Boot and Shoe Makers and Brush Makers. When the Society dissolved Isabella Ford formed a Workwomen's Society in 1888.[34]

By the late 1880s the Leeds clothing industry was being transformed. Unlike London, large factories were established, mechanisation introduced and young single women flocked to Leeds from the surrounding countryside. However, these factories also created an expansion of finishing work which was done by married women at home. This kind of homework proved difficult to organise.[35]

In the late 1880s an upsurge of militancy occurred among unskilled and semi-skilled workers; dockers and gas workers were mobilising along with unorganised women workers, among them the matchgirls at Bryant and May. Lady Dilke, who took over the leadership of the Women's Protective and Provident League in 1886 after Emma Paterson's death, observed of this 'new unionism' that its appeal to sentiment was a strength.

> It has shown that it is possible to organise with effect – if only for a time – the least skilled and most underpaid forms of labour; that is to say, it has touched the very classes amongst which the League is most desirous of promoting combination.[36]

The initial impulse was to embrace the unorganisable and the impact of this extension of unionisation was felt in many parts of the country. However, homeworking presented even greater difficulties than other forms of sweated work.

In 1888 and 1889 Isabella Ford was swept up by strikes involving women in woollen manufactures and machinists in the Leeds tailoring trade. Like the match girls they objected to charges made by the manufacturers, this time for power. Even factory employers tried to unload costs on to their workers. When they resisted the employers used the skilled male cutters and homeworkers to undercut them.[37]

Attempts to organise homeworkers in Liverpool had collapsed by 1888: 'few joined and only 3 or 4 were willing to make any effort . . . it never succeeded in reaching the lowest class of workers, whose lives are too necessitous to sacrifice any present to future good.'[38]

However, in 1890 in Liverpool a minority of the tailoresses managed, by highly visible action, 'brandishing umbrellas and handkerchiefs'[39] while picketing workshops on the main thoroughfares of the city, to force the employers to negotiate over their grievances, low pay and irregular rates and employment. Key factors were the support of middlemen and male tailors in Liverpool and Manchester who refused to scab.[40] The demands were won initially in the large workshops and then in the smaller ones – a pattern of negotiation which continued well into the twentieth century in the northern mechanised clothing industry.

The Gas Labourers and General Labourers Union in which Eleanor Marx played a prominent role recruited women cotton workers in Bristol in 1889.[41] Among the socialists helping to organise were Miriam Daniel and Helena Born. Like Isabella Ford they had been feminists in the Liberal Party before joining the autonomous Socialist Society in the city. Evangelically they turned to the low-paid seamstresses working at home in the countryside around Bristol.

The historian of Bristol socialism, Samson Bryher, observed in 1929:

> An organisation was effected, but the nature of the membership, women who carried home their work, and were thus largely strangers to one another, proved an obstacle so serious that little seems to have been accomplished. It was pioneer work, and the subsequent success of women's labour unions, even in sweated industries, justified the effort.[42]

Social workers outside work could be important in reaching homeworkers. In 1890 Lady Dilke held a successful meeting of matchbox makers in east London: 'The women took a very intelligent interest in the suggestions made to them from the platform with regard to improving their condition.'[43]

Social gatherings were also popular. Three hundred came to tea with Lady Dilke, but organising a society was a problem. They had no time and it was felt that because they worked at home combination seemed impossible.

> The difficulty to be contended against in organising the matchbox makers does not consist in inability on their part to grasp the principles of unionism. The husbands of many of them are members of the Dock Labourers' Union,

> and they listen with intelligent interest to a discussion of the question of combination.[44]

They were not passive victims or roaming vagrants, simply very busy women.

The league continued to plug away with tailoresses, laundresses, bookbinders, shirt and collar makers, dressmakers, milliners and mantle makers, cigar makers and shop assistants.[45] They even attended to the lowest of the low, rag sorters. The 1891 Report stated:

> Rag sorting is decidedly one of the most unpleasant methods to which women resort for earning a living. Bales of rags arrive from the Continent, picked out from the refuse of the dust-heaps, remain tightly packed for a year or more, and then they become an infected mass swarming with vermin, and when opened out are placed before the women whose work consists in sorting out the rags, separating the cotton from the woollen etc. Many faint on setting to work on a fresh opened sack.[46]

The fragmented structure of this kind of work made organising difficult. However, it was not easy even in large units like Ada Nield Chew's clothing factory in Crewe which did contract work for the government. Very few of her fellow workers supported her stand against sweated conditions in 1894. One of them told Eleanor Marx, who urged them to organise at a meeting in Crewe, that this was because they simply would have been sacked.[47] There were plenty of young women who could make clothes for army contracts to take their place.

The difficulties which faced women's trade unionism led to a change of policy. From 1890 the league was trying to combine organisation with pressure for legislation, for example to prevent women working with dangerous substances like white lead.[48]

Isabella Ford gave considerable thought to the form such a combination should take. She argued that educated women could be helpful, but only if they understood grievances 'from the actual worker's own standpoint'[49] and if their research was based on active participation and listening to the women workers in order to learn from them. This differs significantly from Beatrice Webb or from Clementina Black's investigations of homework in which the women workers are rendered passive in the face of thorough but distanced surveys.

Isabella Ford believed involvement at the grassroots was part of a necessary combination of other strategies: 'Legislation is largely the result of organisation and legislation, in its turn, strengthens the hand of the unionists'.[50] However, even Isabella Ford showed a weariness by 1898 about the organisation of homeworkers. Depression and an employers' offensive meant that the optimism of the early 1890s had waned.

> Some people talk grandly about awakening public spirit in these workers and forming them into a strong organisation to resist unjust treatment. This is excellent advice, if it could be carried out, but anyone who has worked at women's trade unions knows only too well how hopeless it is to attempt to

> combine these half-starved, over-worked, broken spirited creatures, toiling in thousands of isolated rooms and trembling lest by any false step they should lose the work that, ill-paid as it is, is often hard to get.[51]

Changes occurred in the organisation of production as mechanisation developed on a larger scale, creating a new kind of worker and heightening the division between those employed in the new industries and the vast pool of homeworkers. Even the organised workers were on the defensive by the late 1890s and this contributed to a feeling of despair about organising the casualised. It is arguable that after the euphoria subsided new unionism, which had apparently opened up the unionisation of casualised women workers, contributed to assumptions about who could and could not be organised. The concentration on large-scale units of production marginalised non-factory workers. Tramping around after scattered workers or small friendly society type women's unions began to seem too time-consuming, especially as so many of these groups had disintegrated in the early 1890s.

The new militant suffrage movement in the 1900s, however, was to interact with trade union organisation among casualised women workers. In 1904 Mary MacArthur revived Emma Paterson's original idea of a general union of working women and formed the National Federation of Women Workers.

So from 1906 there were three middle-class organisations concerned about and actively organising around the conditions of women's work: the Women's Trade Union League, the Scottish Council for Women's Trades and the National Federation of Women Workers.[52] The Labour Party women also established the Women's Labour League and Isabella Ford became a member.[53] Moreover, the Women's Industrial Council was doing investigative work. Feminist ideas had an impact on all these groupings and on the Women's Co-operative Guild, a largely working-class organisation.

Many feminists were campaigning to improve sweated women's working conditions. Also some socialist women were drawn into the suffrage agitation because they believed that if women had the vote they would be able to end sweating as a system of production. In Leeds the combination of socialism and feminism, established by the work of women like Isabella Ford, persisted. A member of the new generation, Ethel Annaken (later Snowden) became convinced of the need to combine political with economic power in order to overcome women's vulnerability at work. Her influence on Philip Snowden who became a leader of the Labour Party was important in gaining an ally for the Labour-suffrage alliance Isabella Ford and other socialist feminist women were struggling to make.[54] This strand of the suffrage movement in contrast to the military discipline of Christabel Pankhurst stressed feminism as a process of self-emancipation. Political action involved developing an inner power and confidence in one's capacities and overcoming passivity.

Impatient with the lack of effect of Labour in Parliament some younger feminists were inclining towards the industrial organisation and direct action of the syndicalist movement. In 1912 Ada Nield Chew wrote with a certain exasperation to *The Freewoman*, a feminist magazine which advocated direct action, sexual liberation and communal living:

> you appear to have a childlike faith in the immediate possibility of organising women workers into trade unions and indicate that herein lies the salvation of womanhood – just as if you had stumbled on something quite new and undiscovered.[55]

She drew on her own experience in a Crewe factory and on twelve years' work for the Women's Trade Union League to argue that in order to get and keep women in trade unions there must be a wider 'awakening', an inward shift in consciousness in which the woman's dependence on the man shifted.

> If there is lying dormant, one spark of latent desire for freedom, for growth, you have some ground to work on, some hope of results. . . . If she wants a vote she is at least alive. . . . It will be a long business, and will need a new generation, bold and reared by freewomen, to leaven the lump of inanity.[56]

Disagreement over the role of the state persisted. During World War One, Mary MacArthur, who had been at the centre of the upsurge of unorganised women like confectionery workers, rag pickers and tin box makers in London in 1911, became increasingly involved in negotiations with the wartime state. In contrast Sylvia Pankhurst, fiercely opposed to the war, was conscious how the East End on the march, because of its proximity to Parliament, could symbolically challenge the state.

Philanthropy and social investigators

Sweated women's work evaded factory legislation and exhausted union organisers but it attracted Victorian philanthropists and social investigators. Their wide-ranging schemes went from ignorant and condescending charity to proposals for regulation through the licensing of workshops or a legally established minimum living wage based on impressive investigations of conditions – through which the investigators themselves were sometimes changed.

One strategy was simply to physically remove casual workers. For example, Sidney Herbert presented an emigration scheme for poor needlewomen which he was convinced would put an end to the slop-work exposed by Henry Mayhew in his articles in *The Morning Chronicle* in 1850. A Fund for Promoting Female Emigration was duly started and a group of needlewomen carefully vetted and selected. *The Economist* criticised Herbert and claimed Irish immigrants were the cause of distress among the needlewomen. Shipping women out to Australia would not solve the problem which was that the market was overstocked. More immigrants would come to England to replace the emigrants. And what were needlewomen to do in Australia? The men there needed women who could run households and farms.[57]

Herbert's scheme was not a great success. It was absolutely true that the needlewomen were incapable of milking cows. Not only were conditions on the ships brutal, there were reports of immorality. Women carefully vetted for respectability mysteriously got pregnant. Emigration schemes in future were careful to send off parties supervised by a matron. But the demand was for domestic or farm servants. Only superior needlewomen were needed so emigration could not tackle the problem of the slop-women.[58]

The Christian Socialists and later Ruskin in the 1870s proposed home colonisation as an alternative to emigration to the colonies. Christian Socialist proposals were on a voluntary basis. However, in the mid-1880s the economist Alfred Marshall suggested that labour encampments should be set up outside London with the co-operation of employers of low-wage labour.[59] The idea of home colonies went through various permutations. Charles Booth thought removal of his 'Class B', the casual labouring class, into regulated colonies where they would work long hours but be fed and housed without pay would solve the problem.[60] There was concern to separate the casual poor from respectable workers and also to distinguish between those who were feckless and those who were criminal. Another advocate of labour colonies Francis Peek wrote *The Workless, the Thriftless and the Worthless* in 1888.[61] The move was technically to be voluntary but Marshall and Booth suggested that it could be effected by making any alternative for the casual poor impossible. He believed this was a 'limited socialism' – by which he meant state regulation of the society.[62] Social imperialists from the 1880s advocated both emigration and home colonies.[63] The evangelical General Booth was also attracted to home colonisation and the Salvation Army actually set up a colony at Hadleigh as a means of rescuing the 'submerged'.[64]

Gareth Stedman Jones shows these to have been voices in the wilderness in the 1880s and 1890s without any influence on state policy.[65] Individualist values were still extremely strong, though during the Boer War, the Interdepartmental Committee on Physical Deterioration suggested labour colonies on the lines of the Salvation Army colony 'with powers however of compulsory detention'.[66] Children of the workers on the colonies would be looked after by the state and the parents would have to work under state supervision in lieu of paying maintenance. The young economist A.C. Pigou wrote in 1900 that such measures appeared to be cruel but to detain 'the wreckage of society' and prevent them from propagating more of their species was in the long run a 'kindness to the race'.[67]

Fear of the inclination of the male casual poor to riot, insurrection and crime was behind the aim to segregate them. This was less a feature of middle-class anxiety about sweated women workers who were seen rather as the reproducers of inferior stock or inclined, because of low wages, to become prostitutes who were not only in need of moral rescue but a social danger because they spread venereal disease. Charles Dickens ran a refuge for prostitutes which was later extended to include 'starving needlewomen of good character'.[68] Providing hostels for destitute women workers was seen as a means of preventing immorality. Homelessness continued to be a problem for women workers and fear of immorality in mixed municipal lodging-houses led to campaigns for women's lodging-houses. Some private women's 'shelters' were also set up.[69]

An alternative approach to these strategies of elimination and segregation was the attempt to regulate the labour market by registering needlewomen. The Association for the Aid and Benefit of Dressmakers and Milliners, founded in 1843, proposed registration. Employers had mixed reactions; some objected to it because they said they could not supervise the morality of the women who were freelance employees.[70] The Distressed Needlewomen's Society maintained the register and provided security for materials. They also trained orphans and 'poor females'. They tried to stimulate the demand for clothing by funding charity orders for the poor and establishing a provident fund for sickness. But these latter two measures had to be rapidly abandoned as inoperable. More modestly they settled for a hostel in Bethnal Green and a campaign against competition from workhouses and prisons which lowered rates.[71]

A Provident Fund was set up by the Milliners and Dressmakers Provident and Benevolent Institution in 1849. It gave free medical attention and made provision for old age and misfortunes. Attempts to form similar philanthropic associations failed in Glasgow and Manchester in the early 1860s but a Female Provident Association was formed in Manchester in 1870. The suffragist Lydia Becker and Manchester Warehousemen and Clerks Provident Association were involved in this early attempt to form links between middle-class feminism and male workers.[72]

The sweated women's lack of marketable skills led to various proposals and schemes for education and training. Engraving or watch-making were suggested as craft alternatives to sewing, and the Christian Socialists set up a Working Women's College.[73]

Towards the end of the century there was a move to revive female cottage crafts which had declined. In the 1880s various philanthropic schemes employed women in the linen industry, handloom weaving, lace making and embroidery.[74] Interest in reviving craft as a means of improving the quality of working life found a left-wing advocate in William Morris, but he saw it as part of a wider socialist alternative to capitalist development which could be combined with factory production. In contrast, other advocates of arts and crafts saw creative, fulfilled work either as an alternative to socialism and militant trade unionism or simply as a means of aesthetic consumption for the rich.

Middle-class arts and crafts groups revived lace making in the 1890s. They improved the standards, conditions and pay of the lace makers who still lived in the Midlands and had faced competition from mechanisation from the 1850s. However, many of these skilled lace makers were between 60 and 90 years old. They found the work a strain and their eyes were failing. The arts and crafts revival was thus rather a mixed blessing to them.[75]

Victorian philanthropy, in an ad hoc way, came to recognise that a combination of strategies was necessary in order to tackle the many-faceted problems of casualised women workers and the boundaries between philanthropy and political action could sometimes become blurred. Middle-class radicals, socialists and feminists all got involved in projects which they thought would improve conditions. Some of these liberal and socialist interventions sought to inspire what R.H. Tawney called 'enlightened discontent'.[76] Others wanted to arouse human compassion like Dickens's Ghost of Christmas Present, by making the conditions visible.

In the 1870s middle-class reformers like Leonard Montefiore and Arnold Toynbee started the settlement Toynbee Hall. Inspired by Ruskin and by a visit to American utopian communities, and fired by metaphors of colonisation of the slums, they went to live and investigate slum conditions, setting up projects and campaigns in east London. Influenced by the study of social science, the 'settlement' idea was taken up in other parts of London and in other industrial cities.

Women joined the settlements and began to study housing and welfare needs. Girls' clubs offered, besides dancing, sewing and millinery. In the early 1890s Emmeline Pethick and Mary Neal formed a small settlement, the Espérance Club. They decided that schemes for improving leisure were not enough and began 'to study the industrial question as it affected the girls' employments, the hours, the wages and the conditions'.[77] Just like the initiators of the register in the 1840s they noted the periods of over-employment and under-employment of women in the West End dress trade because of seasonal demand. To offset overwork and poverty they opened a model workroom in 1897 and employed five young women for forty-five hours a week with a minimum wage of 15 shillings. By the end of the first year they employed twenty. They also taught the girls economics and gave information about trade unions and co-operatives. The arts and crafts revival had an impact too; English folk songs and dances were part of the education.[78]

From the 1880s the social imperialist lobby was seeking anti-alien laws. W.H. Wilkins, the author of *The Alien Invasion* in 1892, believed that foreign immigrants sweated English women and drove them on to the streets. If he had read Charles Dickens's journal *All The Year Round* he would have been able to get an alternative perspective from a homeworker. In her opinion Christian and Jew alike drove hard bargains: 'If there were a shade of preference, perhaps upon the whole she would rather work for Jews, for there was less pretence about them . . . she really thought the Jew was pretty nigh the better Christian.'[79] He might have also met 'her baby – a thin and solemn baby, sitting quite sedately in a very tiny chair, and staring silently at mother while she pursued her work' and a 'curly light-haired' ragged boy with 'sore skin'.[80] If he had, he might have been saved from a painful and disillusioning economic venture. By 1894 his Association for Preventing the Immigration of Destitute Aliens was defunct and Wilkins tried another scheme.[81] He opened a parochial outworkers' room in Bethnal Green with a Reverend W. Davis and urged homeworkers 'to profit by its superior warmth, light and air'. However, it was a failure and he wrote grumpily:

> Our outworkers' room was closed last Saturday. It was so ill attended that I did not find myself justified in continuing it. A few women came, but very

> few. The experiment has once again proved the truth of what we so often feel – viz – that of all people, poor women are the most difficult to help, in as much as they will not play up to help themselves.[82]

The light-airy-room strategy which sought to group workers together and improve conditions was appropriate to young girls and single women but not to married women with children unless there were creches. Also as we have seen it only seems to have been economically viable in certain branches of the trade. Many philanthropists were apparently completely unaware of previous measures and managed to repeat the most obvious mistakes because of their disregard for the actual situation of poor women workers.

In contrast, Charles Booth's *Labour and Life of the People* aimed at an objective assessment of social circumstances as a guide to action. It remains an impressive work – even though Booth had his own subjective and emotional assumptions about the casualised poor. Interestingly he proposed a series of combined strategies for sweated workers.

> What is needed for working women in general is a more practical education in the Board Schools; greater facilities for the exercise of thrift, and definite instruction in the advantages and best methods of saving. If the women and girls will not go to the Post Office Savings Bank, is it quite absurd to suggest that the Post Office Savings Bank should go to the women and girls? And lastly, and not least, trade union is wanted, not union against employers, but union with them; a recognition on the one side of the need and advantage of having good organisers whose exceptional ability makes them worth an exceptional reward; an acknowledgement and acceptance on the other of the responsibility which lies with everyone whose position, talents or advantages have made him his brother's keeper. The question of wages is trivial compared with the question of regularity of employment and kind and just treatment.[83]

Booth imperceptibly drifted into the masculine gender when he wrote of the exceptional characters who would assume this dutiful kind of union stewardship. His support for unions, however, marked a new phase in philanthropy. Unions came to be seen as a form of collective self-help. Thus to organise unions among women especially could be an aspect of philanthropy – another indication that borderlines between strategies were not always clear. Unionisation has not only taken various forms, like co-operative associations, it has had different social implications.

Women investigators followed the rigorous standards pioneered by Booth. In 1894 the Women's Industrial Council was formed evolving from the Women's Trade Union Association. Like the Women's Trade Union League it had close links with the Liberal Party but included socialist women. It was mainly concerned with investigations of women's work and with lobbying for legislation though there were not hard and fast divisions. There was a central council and local groups formed throughout the country.[84] In Liverpool the local group did research on homeworkers in the clothing industry. Eleanor Rathbone, who was later to lead the campaign for child benefit,[85] produced a report on homework in 1908.[86] She also tried to form the Liverpool Association of Homeworkers and Outworkers, but only sixty women were recruited and the attempt failed. She adopted a combination of approaches, organising the Domestic Workers Insurance Society and founding Liverpool University's School of Social Science.[87]

Anxiety about the high infant mortality rate contributed to a revival of concern about the impact of homework on motherhood in the early twentieth century. Clementina Black embarked on an ambitious study of married women's work for the Women's Industrial Council in 1908 with a team of investigators. It was finally published in 1915 and provides a wealth of material about homework. Its weakness is that it portrays homeworkers as passive objects of investigation.[88] The Women's Co-op Guild, the fabians, the Anti-Sweating League and the Women's Labour League were all busy collecting information on sweating in this period.[89] There were links between the various organisations who were investigating sweating. For example, Clementina Black was involved in the suffrage movement and in the Anti-Sweating League. Margaret MacDonald (wife of Ramsay) was involved in the Women's Industrial Council and also the Women's Labour League. Interconnections were not restricted to Britain. Margaret MacDonald collected information from Ireland, as well as Illinois, Massachusetts, Ohio and New York, about methods of regulating sweated work. She was impressed by the American system of licensing small workshops which met the approved health and safety conditions.

Investigation did not, however, always lead to strategic agreement. Margaret MacDonald and Clementina Black were both reformist socialists concerned about women and labour, but were at the centre of a conflict which rocked both the Women's Industrial Council and the Women's Labour League in 1908–9. Clementina Black argued that trades boards to establish minimum wage rates were the answer. Margaret MacDonald took the position of many trade unionists and argued that fixed minimum wages would perpetuate low pay. Mary MacArthur from the National Federation of Women Workers, however, said that licensing was inadequate and argued for trades boards. Margaret MacDonald insisted that her American research showed that licences worked and accused Mary MacArthur of not being 'up-to-date'.[90] In 1908 the Women's Labour League came out in support of trades boards and called for a government bill. The controversy continued.[91]

There had been a significant shift from Booth's argument that irregular employment was the problem. The investigators of the early twentieth century insisted that low pay was the key factor. They believed low-paid women contributed to social and economic problems; expediency and morality fused. At a church conference in 1913, Constance Smith took the long view and set wage rates in an international context:

> In the long run the better article – and the well-paid, well-fed worker will always produce the better article – is certain to beat the worser in the world's market; we

> have seen this happen again and again in the case of industries where the difference lay, not between downright sweating and a living wage, but between decent remuneration and wages that could be called distinctly good.[92]

An economic case for high wages and welfare improvements was thus being presented. According to this scenario better conditions and pay at work meant superior goods and more effective competition with foreign rivals. A healthy educated workforce was a form of investment. Some of the unfit, sick and elderly should be removed from the labour market and supported by the state. As for casualised workers, the young Liberal on the Board of Trade, William Beveridge, along with Winston Churchill, maintained the residuum should be treated as if they were psychologically ill. Beveridge argued they should be deprived of 'citizen rights' and not allowed to father children.[93] As for the mothers, concern about the physical condition of the working class meant that some social investigators opposed married women working. There was a hidden agenda in some aspects of the campaign against sweating – homeworking should not simply be regulated, it should be abolished. Middle-class responses to sweating were not simply moral but strategic. Because the Liberal upper middle-class intelligentsia exercised a disproportionate influence upon state policy, often by informal networking rather than formal positions, by the early twentieth century links formed in settlement projects and campaigns were beginning to have an effect on state policy.

Using the State

The Factory Acts, even in the era of *laissez-faire* in Britain, had established that the state should intervene in the economy to protect women and children. Attempts to extend this protection to workshops and homework in the late 1870s had proved more difficult. While workshops were nominally covered, liberal individualism still found state inspection of the home a violation of personal freedom – and as we have seen, the outworker trade unionist Mrs Mason was also opposed to the state regulating homeworkers.

By the 1880s and 1890s old-style liberal individualism was being assailed by the new social liberalism, fabians, ethical socialists and marxists. All these groups believed the state could be used for the welfare of society as a whole and were prepared to forgo individual rights for what they believed to be the collective good. *Laissez-faire* was also under attack from a populist right calling for state intervention to protect the imperial race from degeneration.

Lord Dunraven, a Tory Democrat, proposed a House of Lords inquiry into sweating in 1888, convinced that it was caused by Jewish subcontractors. Though the other members of the committee showed subcontracting was a symptom not a cause, he persisted in his beliefs and was part of the anti-alien lobby.[94]

A practical innovation in using the power of the state as a big purchaser came from the Progressives, a radical alliance in local government. Local authorities already had some powers to prevent insanitary conditions. In certain trades the employer had to keep and furnish lists of outworkers for the convenience of local authority and factory inspectors. The local authority also had powers to prohibit giving out work. In 1890 Clementina Black, representing the Women's Trade Union League, persuaded the London County Council (LCC) to extend these powers by including women in the clothing trade in its fair wages ruling. The Council had agreed under pressure from trade unions they would only give out work to firms paying decent wages.

Clementina Black explained that this meant the LCC would have to decide what a fair rate was for clothing workers.

> There is no valid recognised Union rate of wages in the ready-made clothing trade, so far as women are concerned; and the rates actually paid, are in many cases, too low for health or reasonable comfort on the part of the workers. This is true of factory workers as well as those employed by middlemen or at home. If the Council desires to pay a fair rate, that is a rate at which the women employed can live in health and reasonable comfort, the Council will have to establish such a rate.[95]

She suggested it should be 4d an hour, thus 17s 2d for a fifty-four-hour week, and was careful to insist that deductions for thread, trimmings, steam power and cleaning should not be made.

Clementina Black's proposal sounds like common sense; in fact it was quite a startling innovation extending local government's responsibility for welfare at work. Not only was the state to decide wage rates, her concept of a 'fair wage' was based on workers' needs: a 'living wage'. It implied that clothing manufacturers were responsible for the health of their workers and that sweated workers had the right to lives of 'reasonable comfort' – an alarming hedonistic concept.

The LCC's committe on contracts decided to ban subcontracting in the firms they dealt with and impose heavy penalties. When the House of Commons passed the 'Fair Wages' resolution in 1891 these terms came to apply to all government contracts. It seemed that state purchasing power was really going to affect those forms of production, particularly the lower grades of clothing, which used subcontracting, small workshops and homeworkers. It also had implications for the workers employed. However, a loophole made subletting acceptable if it was 'inevitable'. Like 'reasonable comfort' this was a term open to several definitions. Also the Fair Wages resolution was difficult to enforce without a strong inspectorate and a conscious trade union organisation which could reveal employers' evasions. Even here there was an ambiguity, for losing a public contract would be bad for the workforce too.

However, a minimum scale of wages was fixed by the LCC and a list kept of 'Fair' and 'Unfair' houses. The concept of the state intervening to determine wages had been established – or perhaps re-established is the better word, for it had existed before the free market had become ascendant. However, it could not be practically applied as a means of raising homeworkers' rates and consequently the LCC banned work going to homeworkers in 1894.

The 1891 Factory Act required the registration of workshops and enabled outworkers to be registered.[96] The number of factory inspectors was increased and a women's department created. There were five inspectors in 1896 and twenty-seven by 1906 but they could still hardly check all the places of women's employment.[97]

The powers of local government to make sanitary inspections of workshops were strengthened in 1895. The snag, as James Macdonald, the Secretary of the Amalgamated Society of Tailors, pointed out, was that the employer simply sent work to another equally insanitary house. 'The only sufferer is the outworker who must leave or lose his work.'[98]

This was a persistent problem with proposals and measures throughout the 1890s. The sweated workers' and homeworkers' interests actually had a very low priority. They were seen as a sanitary or eugenic problem, a threat to trade unionism, as an inefficient system of production, or a threat to public order. For example, Charles Dilke, the Liberal MP who brought a Bill for minimum wages before Parliament every year between 1898 and 1906 to no avail, was convinced that the 'stability of the State'[99] was menaced by the existence of an 'unorganised and depressed body of workers'.[100] Both social liberals like Dilke and ethical socialists regarded the state as an embodiment of a transcendent good. In the face of such lofty concerns the specific complaints of homeworkers and sweatshop workers about paying for thread or power, about travelling to collect materials and deliver work, about tram fares and childcare, problems of credit, the low rate for shirts and the bother they had with buttonholes, appeared trivial and petty.

By the early twentieth century a vast amount of information was gathering dust in Blue Books as a result of government commissions. But two themes came to predominate among the campaigners against sweating: insanitary workplaces and low pay. The Boer War revived anxiety about physical deterioration and social imperialist ideas were again becoming popular, influencing not only the right but also some socialists. Homework was associated with inadequate mothering and concern about sweating increasingly focused on this aspect. There had also been an important change in the attitudes of large employers to pay. George Cadbury for example, the Quaker cocoa manufacturer, was convinced that sweated work was evil. He also believed that while union organisation and better conditions might in the short run threaten profits, it was better to take a long view. Large enlightened companies negotiating with responsible trade unionists were preferable.[101] Moreover, this changing perspective, despite the protests of small employers, also had advocates in the new Liberal government. Campaigners in Britain had influence within the civil service, particularly in the Board of Trade. This political and administrative support within the state made a crucial difference.

George Cadbury agreed to finance an Exhibition of the Sweated Industries. Mary MacArthur enlisted the support of the liberal editor of Cadbury's paper *The Daily News*, A.G. Gardiner, so the exhibition was guaranteed publicity. It opened on 2 May 1906 and fashionable London was devastated to learn that their cigarettes, matchboxes, the beading on ladies' shoes, their stockings, jewel-cases, tennis balls, belts, ties, furniture, brushes and saddles were all being made by homeworkers who after a twelve- to sixteen-hour day earned between 5 and 7 shillings a week.

It was not a completely original idea. A small exhibition had been put on in Bethnal Green in 1903, and in Berlin in 1904 a sweated trades exhibition had made considerable impact. However, it was an impressive propaganda exercise, not only because of the direct encounter between the sweated workers and consumers but also the coverage in the *Daily News* and the detailed catalogue reporting on each trade. The exhibition tugged not only at heart strings but at fear of infection and a backward, inefficient system of production holding back Britain in the world markets. Lily Montague described cigarette makers competing with machinery: 'The workers use a kind of coarse, starch paste, and lick the cases to fasten them. They follow this unpleasant method because any other contrivance for moisturising them would involve greater expenditure of time.'[102]

Gertrude Tuckwell, who wrote the preface to the handbook of the exhibition, had worked closely with Mary MacArthur. After the death of her aunt Lady Dilke in 1904 she had become President of the Women's Trade Union League.[103] She forcefully marshalled the arguments against sweating and homework in particular. It 'pressed . . . into service' a labour force of the unfit, 'the aged and infirm, the crippled and the half-witted'. It forced children to 'toil early and late'.[104] It thus perpetuated a new generation of unhealthy, badly educated casualised workers. 'How shall we put hearts and energy into citizens reared under such conditions?'[105]

Homework not only made women neglect their homes while the multiple use of rooms increased the danger of disease,[106] Gertrude Tuckwell also thought it was an inefficient, parasitic form of industry which held back economic progress which she saw as factory production monitored by 'law, publicity and public opinion',[107] adding: 'Hidden away in the holes and corners of our cities, homework is . . . a survival of different labour conditions, and has no place in healthy, modern industrial life.'[108]

Factory regulation, consumer boycotts, licensing fair shops had all proved inadequate; the solution she advocated was wages boards to fix rates of pay in sweated industries[109] – a remedy Charles Dilke had been pushing for since the 1890s. The Australian Wages Boards introduced in Victoria in 1893 were the model. Although Gertrude Tuckwell said these would 'protect the workers themselves',[110] she obviously believed that homework should be eradicated. She assumed that inefficient workers and married women would be 'shaken out', preventing them rearing a sickly stock.[111] The problem was quite where those who could not rely on an increased male wage were to land, once 'shaken out'. Some opponents of sweating were linking intervention in production to reproduction. The Fabian Women's Group and the Women's Co-operative Guild were arguing for mothers' allowances or pensions. These were originally conceived as sufficient to maintain a woman and her children. Welfare was to be a net for a section of the casual poor.

Having succeeded in making the sweated workers' conditions visible, the exhibition organising group set up the National Anti-Sweating League for a Minimum Wage.[112] Clementina Black's suggestion that they campaign for a 'living wage' was not adopted. Instead the demand was for a minimum defined by 'what the trade could bear',[113] rather than a wage which would meet workers' needs. The league was funded by Cadbury and led by J.J. Mallon.

Later in 1906 a conference was held in London by the league on the minimum wage. It included employers, representatives from the churches, the fabian Sidney Webb, economists like J.A. Hobson, the veteran member of the anti-immigration lobby, Lord Dunraven, and of course Gertrude Tuckwell, Mary MacArthur and Clementina Black. Every species of socialist was there from the Labour Party to the marxist Social Democratic Federation. The league was particularly concerned to convert trades unionists who tended to advocate consumer power and the trade union label. Ben Tillett, the dockers' organiser who had worked with the Women's Trade Union League from the 1890s, told the conference perkily, 'He had a label in his hat; he wore a Trade Union shirt and co-operative clothing.'[114]

James Macdonald, the Secretary of the London Trades Council who had contributed to the 1906 exhibition on tailoring, was also present. Like Tillett he favoured individual and public consumer power buying from better parts of the clothing trade.[115]

Ethel Snowden from Keighley Independent Labour Party (ILP) made an impassioned speech.

> Experience taught her that if they wanted girls to join Trade Unions they must address them on higher ground than wages. They must appeal to the best that is in them. They would go forth from the conference as missionaries to teach others the desirability of supporting a minimum wage, and to spread the doctrine necessary to compel legislation.[116]

The campaign for minimum wage legislation thus was presented as an ethical cause crossing class divisions. She linked abolishing sweating to the suffrage.

> In dealing with sweated girl labour there were problems that would never be settled by them alone, and in giving women the vote they would increase the public spirit of the women of the better classes, who were largely responsible for the evil, and hasten the realisation of their ideals and the solution of the problem.[117]

This was greeted with enthusiastic applause. However, the uncomfortable alliance cobbled together by the Anti-Sweating League looked as if it was going to crack when the Social Democratic Federation declared the wages boards to be a middle-class dodge and demanded the abolition of the capitalist system. Mary MacArthur's graphic and emotional account of sweated women and children toiling in garrets reunited the conference.[118]

Under pressure, the government appointed a House of Commons Select Committee on Homework which reported in 1907–8.119 Mary MacArthur ensured that seven homeworkers gave evidence along with employers and trade unionists.[120] The consensus was in favour of wages boards on the Australian model rather than licensing as advocated by Margaret MacDonald and adopted in Massachusetts. It was thought that the 'casual homeworker' would suffer too much under licensing.[121] The distinction was interesting; it marked a category of decasualised homeworkers.

Members of the Board of Trade, George Askwith, H.J. Tennent, Winston Churchill, William Beveridge, supported compulsory wages boards, though H. Llewellyn-Smith was wary of state intervention; he was finally persuaded. They were careful to avoid a wage level based on subsistence needs and equally careful to explain that the state's intervention in the labour market was not intended to establish a general principle.[122] Employers of cheap male labour in agriculture and the railways were nervous. H.J. Tennent was quite explicit that the Bill was

> to be applied exclusively to exceptionally unhealthy patches of the body politic where the development has been arrested in spite of the growth of the rest of the organism. It is to the morbid and diseased places – to the industrial diphtheritic spots that we should apply the antitoxin of trade boards.[123]

The biological metaphor could have been personal in inspiration as his wife May Abraham, part of the women's inspectorate, had become ill after inspecting sweated conditions. Equally, it indicates the organic view of society and the concept of the state as transcending particular interests which influenced liberal social reform in this period.

The early twentieth-century campaign against sweating was part of a more general change among sections of the liberal ruling class in how they saw industrial relations, the political integration of the working class in capitalism and the creation of a state welfare net. More specifically action on sweating was an alternative to the anti-immigration lobby. Both Dilke and Churchill, along with members of the ILP like Ramsay MacDonald, had fiercely opposed the Aliens Bill when it had been introduced in 1905. However, supporters like Dunraven of the view that sweating was caused by immigrants were also associated with the Anti-Sweating League's campaign.[124] These arguments were not only influencing the right. The socialist trade unionist James Macdonald acknowledged immigration was not the cause but argued nonetheless it was responsible for its rapid development.[125]

The impact of social imperialist arguments about breeding a race to run the Empire was extremely pervasive and was combined with arguments for social reform by some fabians and new liberals. Mary MacArthur cleverly combined moral arguments with social imperialism in a speech at a 1908 demonstration for a living wage for sweated workers.

> It is on behalf of the children that we are here tonight . . . the overworked and undernourished woman is the greatest menace to the prosperity of the nation as a whole . . . we claim that we are the true imperialists [cheers]. No matter

> what our political opinions may be, if we are in earnest in our attempt to solve this problem of sweating, then indeed, we are the true imperialists. It is because we believe with Ruskin that the nation is truly great which has the largest number of happy-hearted children that we are here tonight.[126]

Ruskinian arguments against classical political economy had an important influence on developing concepts of welfare economics from the 1890s and transposed the model of the household upon society. The implications varied. They provided a theoretical legitimation for intervention in capitalism and challenged the inevitability of market forces. They could also lead to a paternalistic view of the state as moral housekeeper. In emphasising an organic concept of the whole society they were inclined to neglect the interests and rights of specific groups, to which both classical liberalism and marxism in different ways were more alert.

Only those aspects of the homeworkers' situation which fitted the overall ideological picture are really taken into account by reformers. The homeworkers assume two roles: they are either starving, completely passive victims aimlessly reproducing a casualised 'residuum' or painfully industrious candidates for decasualisation and regulation. They are not people with views about the economy and society who could be expected to think or do anything about their lives.

It was not the same impression given by Mayhew's slop-woman in the early 1850s, Mrs Mason, the outworkers' organiser in the late 1870s, or even the more sentimental portrayal of the urban homeworker in Dickens's *All The Year Round* in 1884. The ex-factory workers whom Charles Booth described conferring on wage rates for brushes in 1889 also suggested that not all homeworkers were incapable of organising networks. The turn to the state thus coincided with a categorisation of casualised homeworkers as outside the ranks of organised labour. They work but are not really 'workers'. This view was shared by the middle-class reformers and trade unionists.

Two assumptions then were behind the pressure for state intervention in the 1900s; one was that homework is an evil which should be abolished or reduced, another was that homeworkers could not organise for themselves. Either there should be action by a paternalist Ruskinian state or, rather more democratically, the political extension of citizenship to women would enable the upper and middle classes to moralise the state through supposed womanly values.

Only the occasional voice was raised to explain why some workers preferred outwork to factory work. Though James Macdonald was opposed to sweating as a system of production, his experience of tailoring led him to observe that there *were* certain benefits for workers in the system of subcontracting to workshops rented by individuals or groups of workers.

> They can scamp a little of the work, not being constantly under the eye of the employer; [they] may, by working for more firms than one, have a more constant supply of work, and can make a little profit out of employing other hands to help.[127]

The homeworker also shared these advantages and could use family labour. They could 'work as long as they like and save the expense of workshop rent'.[128]

These 'advantages' of course existed within a very narrow span of options and in general circumstances which were horrifying to middle-class observers. On the other hand they make the workers' situation intelligible – especially when the practical need to combine earning with childcare is considered.

Gertrude Tuckwell told the 1907 Select Committee on Homework that it was 'absolutely impossible' to organise homeworkers.[129] An attempt made to organise the tailoring outworkers had failed because they were 'depressed by working night and day',[130] they could not act 'corporately',[131] and were 'entirely scattered'.[132] She had consequently resolved to concentrate the Women's Trade Union League's efforts on enforcing and developing new legislation. Clementina Black shared her view.

However, Mary MacArthur insisted in her evidence to the Select Committee that homeworkers could not be artificially abstracted from women workers in factories. She said: 'the low rates of wages are not confined to homeworkers, and that the question of organisation is equally difficult with the similar class of labour in the factory. Almost equally difficult when the wages are very low'.[133]

Organisation was difficult then for casualised women workers as a whole. But it was not, as we have seen, impossible. Indeed in 1907–8 some organisation of homeworkers in London existed. A police court missionary had started a Homeworkers' Aid Association with 1,000 members, working for their own living and not supplementing the wage of husbands and fathers, without subsidies. He told the Select Committee in 1907 that he favoured wages boards *as well* as organisation because he believed they would raise wages of homeworkers.[134] The Anti-Sweating League had also set up a Homeworkers' League. They opened a Homeworkers' Hall in Bethnal Green to promote association and contact among homeworkers. A witness before the Select Committee in 1908, Miss Vynne, confirmed the League had a membership of around 3,000. The Homeworkers' League was opposed to wages boards but not against state intervention.[135]

Though there were strong pressures to abandon self-organisation for state intervention, there were countervailing economic and political forces. It is tempting in retrospect to conclude that because something happened it was inevitable. But the shift towards relying on the wages boards was not preordained. It institutionalised a specific ideological approach which because of a particular convergence of pressures from without and within the state triumphed as a means of regulating low-paid casualised labour.

The opposition was not united, ranging from advocates of licensing or other measures, the critics of the wages boards in Australia and New Zealand, to the anti-alien lobby on the right and those socialists on the left who were uneasy about bureaucratic collectivist solutions. Though subsequently to be marginalised, they were still arguing when the boards came into force in 1910.

Not all supporters of wages boards and state intervention, however, saw these as an *alternative* to self-organisation. Both feminists, who stressed the importance of the suffrage, and women trade unionists, put various degrees of emphasis upon the *combination* of mobilisation of casual women workers and state intervention. For example Florence Exten-Hann, a young activist in the shop workers union, a socialist and feminist from Southampton, was active in the Anti-Sweating League and in the Women's Social and Political Union.[136] Margery Corbett-Ashby, a constitutional suffragist and liberal from a Scottish aristocratic family, marched proudly with factory workers and chain makers and helped other suffrage campaigners in publicising the wages and conditions of sweated women workers. They exerted influence on male politicians even without the vote. 'We got the facts and publicised the facts. There were men in the House ready to pick up on what we were doing.'[137] Consequently when Mary MacArthur and Julia Varley organised Cradley Heath chain makers they could rely on support from feminists as varied in class background and politics as Florence Exten-Hann and Margery Corbett-Ashby.

In January 1910 the boards came into force in four trades, chain making, lace making, paper-box making, wholesale and bespoke tailoring – they applied only in England, not to the United Kingdom as a whole.

Their effect varied. It was dramatic in chain making at Cradley Heath near Birmingham. Women made chains in their own homes while most of the men worked in factories. Cradley Heath outworkers had a history of organisation. They had taken action in the 1850s but the union had petered out when their claims were won. Attempts to organise the women had been made again in 1886 and the Women's Trade Union League had been involved in campaigning with feminists for the women chain makers.

So when Mary MacArthur and Julia Varley, an ex-textile worker, arrived to set up the National Federation of Women Workers there was a background of organising. Also the chain makers were a local fixed labour force which could not easily be replaced even though they worked at home. They mobilised demonstrations, raised funds and publicised the trades boards among the chain makers. After minimum wages were introduced the employers locked out about 800 women. The National Federation of Women Workers provided 4 shillings a week strike pay. Twelve of the women went to London to collect money and the TUC, feminists and socialists gave support. The press supported the women workers and they won the increased rates set by the trades boards.[138] Their victory inspired other women working in 'hollow-ware' in the metal trades and also women brick makers in the Birmingham area.[139]

Lace finishers in Nottingham presented more problems. Unlike the chain makers their husbands were not in regular employment. J.J. Mallon, Secretary of the Anti-Sweating League, described them as 'hawkers, casual labourers or ne'er-do-wells'.[140] The lace finishers had about 700 women 'intermediaries'.[141] Some of these were 'just and fair dealing',[142] others were extorting 'as much as 50% of the total payment made by the warehouse'.[143] J.J. Mallon also notes that the lace finishers had got into debt. Some of the middlewomen

> had become the lenders of minute sums of money to their workers, on which they charged heavy interest, while others were the proprietors of small shops where, it is said, their employees were compelled to deal, but where they often failed to get full value for their money.[144]

Trade was not going well. In the face of foreign competition the trade board set the rates fairly low.[145]

Among paper-box makers there was a certain hierarchy – those making fancy boxes were the best paid. The trade was partially mechanised and homeworkers got the same rates as factory workers; however, they had to provide their own glue and paste. Like lace finishing, foreign competition was affecting the trade and again the board rate was low, though a little more than they had earned before.[146]

The existence of boards had an influence on the organisation of paper-box making. In 1913 J.J. Mallon said there was better training for workers unable to keep up with the speed necessary to earn the rate. One employer had eliminated the time spent by workers in waiting for work. Now there was 'steady and continuous employment of the workers'.[147] The employers were stimulated by the boards to overhaul production methods and equipment. Many who had been originally hostile to the Act were now in favour.

Unionisation among clothing workers in workshops and factories increased after the Act. But homeworkers' pay did not significantly improve. In clothing the Tailoring Trade Board fixed only minimum-time rates and left piece-rates to the employer.[148] Rates were set at differing amounts for men and women. They were based on estimates of what an 'ordinary' worker could do on time rates. Sylvia Pankhurst wrote to the *Daily Herald* in 1912 pointing out that while men got 6d an hour 'an ordinary' woman worker got 3¼d. Moreover, the employers were simply giving women an extra workload. One of the women workers told Sylvia Pankhurst: 'the outworkers are finding the regulation minimum wage is evaded by employers with impunity and that garments are given out to the women who get threepence-farthing, which they find take not an hour, but three hours to finish.'[149]

When in 1913 six working women from east London went on a suffrage delegation to Prime Minister Asquith, they explained the precise problems of needlewomen, cigarette packers, shoe makers and brush makers. They had a much deeper effect upon him than middle-class campaigners and he promised to consider their case that economic reform could not be achieved unless women had a voice in choosing representatives in Parliament.[150]

Sylvia Pankhurst's aims in organising in east London are summarised in *The Suffragette Movement*:

> I wanted the women of the submerged mass to be, not merely the argument of more fortunate people, but to be fighters on their own account, despising mere platitudes and catchcries, revolting against the hideous conditions about them, and demanding for themselves and their families a full share of the benefits of civilisation and progress.[151]

Sylvia Pankhurst's brand of socialist feminism quite explicitly opposed the view of the casualised women workers as unorganisable – though the minutes of the East London Federation of the Suffragettes show it was far from easy to build a sustained group campaigning locally.[152] She did not approach the state as an 'IT' to be used or captured but as controlling resources which by grassroots mobilisation at work and in communities were to be claimed by workers – including women who were housewives and homeworkers.

During World War One statist and anti-statist strands in socialism and feminism polarised. If Mary MacArthur symbolised the former, Sylvia Pankhurst expressed the politics of the anti-statist left.

But the relationship between the state and production was changing. The Liberals and Labour members in the wartime coalition meanwhile drew on earlier developments of state welfare and intervention in industry. The basis was being laid for the regulated economy and the welfare state. During World War One the structure of women's employment became more integrated within the formal sector, though, in the 1920s when unemployment returned, many women went back to domestic service or casualised work.

The late nineteenth- and early twentieth-century enthusiasts for state regulation imagined that small 'sweat-shops' and homework would die out. Instead they persisted alongside the growth of both Fordist-style factory production and public services, while the invisible nature, particularly of homework, meant that it existed outside official economic statistics.

Although the idea of *combining* organisation with legislation persisted in relation to the organisation of women factory workers between the wars it appears to have been abandoned in relation to homeworkers. The actual grievances of homeworkers and their organisation were neglected until the 1970s. Although the trades boards were extended and provided some regulation of wages in low-paid industries, it seems that most investigators and reformers assumed either that they were sufficient or that a minimum wage should be introduced nationally.

Not only was the democratic participation of women in this extreme form of casualised work in shaping alternatives obscured. The emphasis on pay and sanitary conditions with too broad brushstrokes obliterated the seething, intricate differences of circumstances and the specific grievances which the women would sometimes be more ready to voice than their rates of pay: time taken getting to collect materials, unfair fault finding, payment for materials, problems of credit, dangerous substances which harmed children.

The reliance on state intervention as a means of tackling 'sweating' was not a preordained commonsense response. It appeared because of a wider change in views on the economy and industrial relations among influential liberals and socialists who had a formative influence on the Labour Party. They gave the state unusual powers to intervene in the labour market in a period of unprecedented militancy among workers – including casualised women. The nastier bits of capitalism were to be sanitised by state regulation and a bargain struck with labour. Opposing strategies were politically defeated on several fronts.

After World War Two the gains achieved by this regulated capitalism were to be considerable in terms of improved conditions of life and work. However, there were to be losses as well. People who did not fit into the new categories of regular employment inhabited a twilight economic zone. It was as if they were not really workers. Invisible in the present, their past was constructed as if they had been passive victims. The accumulated understandings of trial and error which had been gathered through diverse schemes to change conditions had no coherent, continuous record. Skilled craft workers had their histories, so did male casualised trades; but casualised women homeworkers were simply the objects of social investigation.

Examining the economic and social situation of casualised women workers in the late nineteenth and early twentieth centuries leads to wider questions: the role and purpose of co-operatives and trades unions; the relation of the state to the economy; how wages should be determined; how to estimate poverty, define needs; whether a universal standard of life could be established and whether the good life is simply material or also involves spiritual values. These were being earnestly discussed by contemporaries in relation to sweated women workers. They have an uncanny resemblance to many of the debates now about the development of the Third World and working women's role in this process.

A problem both then and now is how groups like poor women can gain access to the resources and power of the state, without allowing more privileged interests to define how these are applied. Nineteenth- and early twentieth-century philanthropists and reformers were often more concerned with how women *should* work and associate than with the specific grievances which poor women themselves expressed. Nonetheless, there *were* many examples of initiatives which resemble SEWA's multifaceted forms of organising. Some of these combined mobilising at the base, economic bargaining, social services, with political pressure both locally and nationally on the state. These schemes and dreams of alternatives do not have the force of inevitable historical development on their side, for they did not prevail. Their recording does, however, uncover evidence that something better for poor women was and is conceivable.

NOTES

1 See Andrew Sayer and Richard Walker, *The New Social Economy: Reworking the Division of Labour*, Basil Blackwell, Oxford, 1992.
2 See Bridget Hill, *Women, Work and Sexual Politics in Eighteenth-century England*, Basil Blackwell, Oxford, 1989; Sally Alexander, *Women's Work in Nineteenth-century London, A Study of the Years 1820–50,* The Journeyman Press, London, 1983; and Shelley Pennington and Belinda Westover, *A Hidden Workforce, Homeworkers in England 1850–1985*, Macmillan Education, London, 1989.
3 Charles Booth (ed.) *Labour and Life of the People, East London*, Williams & Norgate, Edinburgh, 1889, p. 454.
4 Karl Marx, *Capital, Volume I, Re-action of the Factory System on Manufacture and Domestic Industries*, ed. Dona Torr, George Allen & Unwin Ltd, London, 1957, p. 465.

5 Beatrice Potter, House of Lords Select Committee, *First Report and Minutes of Evidence*, 1888, quoted in Jenny Morris, *Women Workers and the Sweated Trades, The Origins of Minimum Wage Legislation*, Gower, Aldershot, 1986, p. 20.
6 Catherine Webb, *Industrial Co-operation, The Story of a Peaceful Revolution, The Co-operative Union*, Manchester, 1931, p. 75.
7 E.P. Thompson and Eileen Yeo, *The Unknown Mayhew*, Penguin, London, 1973, p. 31.
8 Wanda F. Neff, *Victorian Working Women, An Historical and Literary Study of Women in British Industries and Professions, 1832–1850*, Frank Cass, London, 1966, p. 142.
9 ibid.
10 Christina Walkley, *The Ghost in the Looking Glass, The Victorian Seamstress*, Peter Owen, London, 1981, p. 103–5.
11 ibid., pp. 104–5.
12 Quoted in ibid., p. 106.
13 *Commonweal*, 25 October 1890.
14 *Commonweal*, 1 November 1890.
15 *Commonweal*, 1 November 1890.
16 Women's Trade Union League, *Quarterly Report and Review*, 1, April 1891, p. 9.
17 Women's Trade Union League, *Quarterly Report and Review*, 2, July 1891, p. 15.
18 Eleanor Marx, 'A Women's Trade Union', in Hal Draper and Anne G. Lipow (eds) *Marxist Women versus Bourgeois Feminism*, Socialist Register, Merlin Press, London, 1976, pp. 223–4.
19 ibid., p. 224.
20 See E.P. Thompson, *William Morris, Romantic to Revolutionary*, Merlin Press, London, 1977, pp. 576–9.
21 Beatrice Potter, *The Co-operative Movement in Great Britain*, Swan Sonnenschein, London, 1893, pp. 225–6.
22 ibid., p. 226.
23 Stephen Yeo, 'Towards "Making Form of More Moment than Spirit", Further Thoughts on Labour, Socialism and the New Life from the late 1890s to the Present', in J.A. Jowitt and R.K.S. Taylor (eds) *Bradford, 1890–1914, The Cradle of the Independent Labour Party*, University of Leeds, Department of Adult Education and Extra-Mural Studies, Bradford Centre Occasional Papers, no.2, 1980, p. 85.
24 Isabella Ford, *Labour Leader*, 2 November 1906, quoted in June Hannam, *Isabella Ford*, Basil Blackwell, Oxford, 1989, p. 121.
25 *The Syndicalist*, II (3, 4), March–April 1913, p. 6.
26 See Sylvia Pankhurst Collection ELFS, Institute of Social History, Amsterdam, and Barbara Winslow, 'Sylvia Pankhurst (1905–24), Suffragette and Communist', University of Washington thesis, 1989, pp. 241–2.
27 Sheila Lewenhak, *Women and Trade Unions, An Outline History of Women in the British Trade Union Movement*, Ernest Benn, London, 1977, p. 48.
28 Sidney and Beatrice Webb, *The History of Trade Unionism, 1666–1920*, printed by the authors for the students of the Workers Educational Association, London, 1919, pp. 336–7.
29 Harold Goldman, *Emma Paterson*, Lawrence & Wishart, London, 1974, pp. 44–5.
30 Lewenhak, *Women and Trade Unions*, p. 73.
31 ibid., p. 74.
32 ibid.
33 Hannam, *Isabella Ford*, p. 33.
34 ibid.
35 ibid.

36 Lady Dilke, 'Trade Unions for Women', *North American Review*, 1892, p. 3, quoted in Pennington and Westover, *A Hidden Workforce*, p. 115.
37 Hannam, *Isabella Ford*, p. 40.
38 *Liverpool Review*, 2 June 1888, quoted in Linda Grant, 'Women's Work and Trade Unionism in Liverpool, 1890–1914, Women and the Labour Movement', *North West Labour History Society Bulletin* 7, 1980–1, p. 69.
39 ibid.
40 ibid., pp. 69–70.
41 Yvonne Kapp, *Eleanor Marx*, Vol. II, Lawrence & Wishart, London, 1976, p. 348.
42 Samson Bryher, *An Account of the Labour and Socialist Movement in Bristol*, Bristol, 1929, p. 7.
43 Women's Protective and Provident League, *16th Annual Report*, July 1890, p. 10.
44 ibid.
45 See Women's Trade Union League, *Quarterly Report and Review*, 1, 2, 1891.
46 Women's Trade Union League, *Quarterly Report and Review*, 1, 2, 1891, p. 14.
47 Doris Nield Chew (ed.) *Ada Nield Chew*, Virago, London, 1982, p. 17.
48 See Women's Trade Union League, *Quarterly Report and Review*, 1891.
49 Quoted in Hannam, *Isabella Ford*, p. 51.
50 Isabella Ford, *Women as Factory Inspectors and Certifying Surgeons*, Women's Co-operative Guild Pamphlet, 1898, pp. 7–8.
51 ibid., p. 4.
52 Lewenhak, *Women and Trade Unions*, pp. 114–18.
53 Christine Collette, *For Labour and For Women*, Manchester University Press, Manchester, 1989, p. 35.
54 June Hannam, 'Women in the West Riding ILP', in Jane Rendall (ed.) *Equal or Different, Women's Politics 1800–1914*, Basil Blackwell, Oxford, 1987, p. 229.
55 Ada Nield Chew, 'Let the Women Be Alive', in Chew, *Ada Nield Chew*, p. 237.
56 ibid., p. 238.
57 Walkley, *The Ghost in the Looking Glass*, pp. 112–14.
58 ibid., pp. 114–21.
59 Gareth Stedman Jones, *Outcast London*, Clarendon Press, Oxford, 1971, p. 304.
60 ibid., p. 307.
61 ibid., p. 312.
62 ibid., p. 307.
63 ibid., p. 308.
64 ibid., pp. 311–12.
65 ibid., p. 311.
66 ibid., p. 331
67 Quoted in ibid., p. 332.
68 Quoted in Walkley, *The Ghost in the Looking Glass*, pp. 101–2.
69 See Mary Higgs and Edward E. Hayward, *Where Shall She Live? The Homelessness of the Woman Worker*, P. S. King & Son, London, 1910.
70 Walkley, *The Ghost in the Looking Glass*, pp. 93–4.
71 ibid.
72 ibid., p. 100–1.
73 Neff, *Victorian Working Women*, p. 143.
74 Anthea Callen, *Angel in the Studio, Women in the Arts and Crafts Movement, 1870–1914*, Astragel Books, London, 1979, pp. 2–5.
75 ibid., pp. 142–3.
76 R.H. Tawney, *The Labour Movement and the W.E.A.*, Rochdale Education Guild Notes, item 14, newspaper cutting, *Rochdale Observer*, 28 May 1910, in Rochdale Public Library.

77 Martha Vicimus, *Women, Work and Community for Single Women, 1850–1920*, Virago, London, 1985, p. 233.
78 ibid., pp. 233–4.
79 Anon, 'Travels in the East Part IV', *All The Year Round*, 22 March 1884, in *One Dinner a Week and Travels in The East, An Account of East End Life first published in 1884*, The London Cottage Mission, reprinted Peter Marcan Publications, High Wycombe, 1987, p. 67.
80 ibid., p. 63.
81 Bernard Gainer, *The Alien Invasion*, Heinemann, London, 1972, p. 63.
82 Pennington and Westover, *A Hidden Workforce*, p. 100.
83 Booth, *Labour and Life of the People, East London*, p. 476.
84 Lewenhak, *Women and Trade Unions*, p. 118.
85 Eleanor Rathbone, *The Disinherited Family*, with an introductory essay by Suzie Fleming, Falling Wall Press, Bristol, 1986, p. 10.
86 Grant, 'Women's Work and Trade Unionism in Liverpool, 1890–1914', p. 72.
87 Rathbone, *The Disinherited Family*, p. 10.
88 See Clementina Black (ed.) *Married Women's Work*, Virago, London, 1983.
89 See for example, B.L. Hutchins, *Homework and Sweating*, The Fabian Society, London, 1907 (pamphlet); Margaret H. Irwin, *Homework in Ireland*, Report of an Inquiry, Scottish Council for Women's Trades Union for the Abolition of Sweating, 1909 (pamphlet); Miss March Phillips, *Evils of Homework*, Women's Co-operative Guild, Manchester, 1902 (pamphlet).
90 Collette, *For Labour and For Women*, p. 118.
91 ibid., p. 118–19.
92 Miss Constance Smith, 'Wage Movements in Other Countries', in *A Series of Lectures on the Industrial Unrest and the Living Wage, the Inter-Denominational Summer School, June 1913*, P. S. King, London, 1914, p. 157.
93 Stedman Jones, *Outcast London*, p. 335.
94 Morris, *Women Workers and the Sweated Trades*, p. 8.
95 London County Council Special Committee on Contracts, Inquiry into the Condition of the Clothing Trade, 12 December 1890, in Rodney Mace (ed) *Taking Stock, A Documentary History of the Greater London Council's Supplies Department, Celebrating Seventy-five Years of Working for London*, GLC, London, 1985, pp. 27–9.
96 Morris, *Women Workers and the Sweated Trades*, p. 172.
97 ibid., pp. 174–6.
98 ibid., p. 128.
99 ibid., p. 142.
100 ibid.
101 ibid., pp. 196–7.
102 Lily H. Montague, 'Notes on the Manufacture of Cigarette Cases', in *Handbook of the Daily News Sweated Industries Exhibition, May 1906*, London, 1906, p. 77.
103 Lewenhak, *Women and Trade Unions*, p. 115.
104 Gertrude Tuckwell, Preface, in *Handbook of the Daily News Sweated Indusries Exhibition*, p. 13.
105 ibid.
106 ibid., p. 14.
107 ibid., p. 17.
108 ibid., p. 18.
109 ibid., p. 16.
110 ibid., p. 14.
111 Morris, *Women Workers and the Sweated Trades*, p. 142.
112 ibid., p. 200.

113 ibid., p. 199.
114 National Anti-Sweating League, *Report of 1906 Conference on a Minimum Wage*, London, 1907, p. 20.
115 James Macdonald, 'Sweating in the Tailoring Trade', in *Handbook of the Daily News Sweated Industries Exhibition*, p. 87.
116 Ethel Snowden, National Anti-Sweating League, *Report of 1906 Conference on a Minimum Wage*, p. 60.
117 ibid.
118 Lewenhak, *Women and Trade Unions*, p. 121.
119 ibid.
120 Morris, *Women Workers and the Sweated Trades*, p. 201.
121 ibid., p. 207.
122 ibid., p. 208.
123 ibid., pp. 210–11.
124 Gainer, *The Alien Invasion*, pp. 146–7.
125 Macdonald, 'Sweating in the Tailoring Trade', p. 88.
126 National Anti-Sweating League, 'Living Wage for Sweated Workers, Report of the Great National Demonstration, 28 January 1908', quoted in Morris, *Women Workers and the Sweated Trades*, p. 202.
127 Macdonald, 'Sweating in the Tailoring Trade', p. 87.
128 ibid.
129 Pennington and Westover, *A Hidden Workforce*, p. 116.
130 ibid.
131 ibid.
132 ibid.
133 Lewenhak, *Women and Trade Unions*, p. 126.
134 Morris, *Women Workers and the Sweated Trades*, p. 207.
135 Pennington and Westover, *A Hidden Workforce*, p. 121.
136 Sheila Rowbotham, 'Florence Exten-Hann Socialist and Feminist, Interview 1972', in Sheila Rowbotham, *Dreams and Dilemmas*, Virago, London, 1983, p. 226.
137 Sheila Rowbotham, Mss, interview with Dame Margery Corbett-Ashby, 1970.
138 Lewenhak, *Women and Trade Unions*, pp. 122–5.
139 J.J. Mallon, 'The Minimum Wage', in *A Series of Lectures on the Industrial Unrest and the Living Wage*, p. 148.
140 ibid.
141 ibid.
142 ibid.
143 ibid.
144 ibid., pp. 148–9.
145 ibid., p. 149.
146 ibid., p. 146, 150.
147 ibid., p. 155.
148 ibid., pp. 151–3; see also Pennington and Westover, *A Hidden Workforce*, p. 111.
149 Barbara Winslow, 'Sylvia Pankhurst (1905–24) Suffragette and Communist', p. 99.
150 David Mitchell, *Women on the Warpath*, Cape, London, 1966, p. 273.
151 Sylvia Pankhurst, *The Suffragette Movement 1931*, quoted in Winslow, 'Sylvia Pankhurst (1905–24) Suffragette and Communist', p. 113.
152 See East London Federation of the Suffragettes (ELFS) Minutes, Institute of Social History, Amsterdam.

Chapter 8

Homework in West Yorkshire

Jane Tate

ABSTRACT

During the 1970s in Britain, feminist, community and research groups were becoming aware of homework as a low-paid and often dangerous form of women's work. The 1980s saw the formation of organisations which have used several methods to publicise and organise homeworkers.

By the late 1980s these had started networking internationally, not only in Europe where the Common Market makes homework a policy matter but also in the Third World. SEWA's approach (see Chapter 5) has been an important influence upon them. Information is now being gathered through international organisations like the ILO which help to make visible this hidden form of labour.

Homework, however, takes many forms as Jane Tate shows in this study of one region, West Yorkshire. Her account assesses the varying strategies used by the West Yorkshire Homeworking Group and also outlines current international debates on homework policies.

THE BACKGROUND

The West Yorkshire Homeworking Group is part of a network of groups around the country which has grown up since homework was revived as an issue in the 1970s. Their concern is not only to document homework but also to bring about change, in particular legal measures to protect homeworkers. A comparison with the agitation early in the twentieth century shows both continuity and change. The most important parallel is the part played by the women's movement and feminist ideas in general, in terms of both research and organising. The two periods have seen an alliance of forces, with women's organisations co-operating with trade unions, researchers, sections of the church and others. But there are also important differences arising from the specific historical context. First, there has been a change of emphasis in the groups working on homeworking, from acting on behalf of homeworkers to working with them. Second, the position of black and Asian women in Britain today, and their role in the homework campaigns, are of crucial importance. The third main difference is the international outlook of the campaigns which have begun to understand the links between women around the world as the basis for solidarity and mutual education.

Before looking at the development of the West Yorkshire Homeworking Group in detail, I will outline the developments of the 1970s and 1980s in relation to homeworking which laid the basis for our work.

Campaigning in Britain from the 1970s

The question of homework re-emerged in Britain in the 1970s. There is no doubt that homework itself never disappeared from the industrial scene, particularly in the clothing and leather industries.[1] In World War Two, home workshops were set up and women were encouraged to contribute to the war effort with such work as assembly of parts for aircraft.[2] While the trade unions continued to express concern about the extent of homework during the years following the big campaigns in the early part of the century, it was not until the rise of the women's liberation movement in the 1960s that the issue of homework became more widely noticed again. Compared to earlier feminist and women's organisations, the recent movement put more emphasis on the relation between paid employment and domestic work in the family. In relation to homework, it showed how in spite of the huge expansion in married women's paid employment, particularly in part-time employment since World War Two, the lack of childcare was still forcing many women into badly paid homework.

In the 1970s, other organisations also took up the question of homework. In 1974, Marie Brown wrote *Sweated Labour: A Study of Homework* as a result of an appeal for information from homeworkers on a popular radio show.[3] It revealed shocking rates of pay, some of them illegal. This was followed up by further work by the Low Pay Unit and the Fabian Society arguing for the need for regulation of homework and specifically a Bill to protect homeworkers. Attempts were made to set up a newspaper for homeworkers but this seems to have been too ambitious and did not last long.

At national level, in 1978, the Frank White Bill was put before Parliament as a Private Member's Bill but failed to get support. If it had become law, it would have given homeworkers the same rights as other employees. The publicity and campaigning of the 1970s did lead to a Select Committee, whose limited recommendations were promptly ignored. It also led to considerable discussion within the trade unions around the exploitation of homeworkers and the need for regulation.

In the 1980s, with the change in government and a decade of deregulation of employment and weakening of the trade unions, the emphasis changed to organisation, particularly at the local level. However, the 1980s also saw the growth of national organisations, in the form of national conferences and later the National Group on Homeworking, an alliance of local groups, researchers, trade unionists and others concerned with the issue of homeworking.

The first local homeworking group was formed in Leicester, where there had been a hosiery and knitwear industry for generations, and was funded by the local authority. Since the existence of homework in the area was widely acknowledged, support for the campaign was broad-based, and came through the Environmental Health Department, a recognition of the importance of the issue of health and safety for homeworkers. The Leicester Outwork Campaign was strongly influenced by relatively new approaches to community work which emphasised helping groups develop self-organisation rather than relying on preconceived forms of trade union organisation. Campaigners from Leicester stressed from the start the need to listen to, and take seriously, the views of homeworkers themselves rather than deciding in advance what was good for them. Feminism has also had a strong influence; links between work and family, or wages and health, training and childcare have been acknowledged as important.

The city of Leicester is now probably the most important Gujarati settlement in Britain, and as elsewhere many Asian women in Leicester work at home in the clothing and hosiery industry. As the campaign developed, it saw the need to work with and employ Asian women. Its outreach workers make contacts with homeworkers in both rural and city areas. Besides advice and support, training courses have been run to meet the demand for improved skills for women, to enable them to get better jobs, either at home or outside, and legal cases have been fought over the right to redundancy pay, for example. At the same time, the campaign has supported other local groups and played an active national role. Before the setting up of the National Homeworking Unit in 1987, the Leicester Campaign and the London Homeworking Campaign were the main groups involving homeworkers nationally. Increased publicity and activity round the issue of homeworking resulted in the National Homeworkers' Charter in 1984 and a statement on homeworking by the Trades Union Congress in 1985. By the end of the 1980s both the Labour Party and trade unions were looking seriously at legislation to give homeworkers employment rights. At the local level, campaigns had developed in spite of funding difficulties and in some areas Labour councils had appointed local authority officers to work with homeworkers.

In the same period, a change occurred in the attitude of many trade unions to women in general and concern mounted about increased flexibilisation or casualisation of the workforce. Within this framework, homework can be understood not as an unusual throwback to Victorian times but as the most flexible form of labour in a whole spectrum of part-time, casual and temporary employment relationships. While most unions have had too many problems on their hands over the last ten years to take much practical action on the issue of homeworking, more attention is being given to the specific needs of women and more attempts made to build appropriate organisational forms. Alliances have also been made with voluntary organisations to campaign against the abolition of the Wages Councils, as well as closer links with the homework campaigns.

These developments coincided with a growing awareness of the international nature of the economy. The same giant companies that employ thousands of electronics workers in the Third World may also be employing homeworkers here. Similarly, the clothing retailers who source their supplies in the Philippines or Thailand may also subcontract work within Britain and sell goods made by homeworkers here.[4] The impact of the European Community on Britain is another factor. Concern about homework led to a study by the Council of Europe, published in 1989,[5] while the Self-Employed Women's Association (SEWA) in India lobbied the International Labour Organisation to hold its expert meeting on homework in October 1990. There have also been growing links between organisations in different countries and an awareness of the need to learn from each other. For example, in November 1992 the International Ladies' Garment Workers' Union (ILGWU) called an international conference to discuss the organisation of homeworkers in Toronto.

As a result of the different approach to homework over the last twenty years, the campaigns are now much more clearly based around the needs and interests of homeworkers themselves, instead of advocating a ban on homework or other legislative solutions. This has meant that the central demand for employment rights for homeworkers has been linked to a variety of other social demands and integrated local, national and international perspectives. The emphasis on working *with* homeworkers, by involving them in campaigns and providing practical services for them, has also brought into focus the different aspects of women's lives, in particular the interrelationship of class, race and gender on homeworkers in Britain today.

WEST YORKSHIRE HOMEWORKING GROUP

Research and contacts

The West Yorkshire Homeworking Group provides an example of a non-traditional way of working with homeworkers, a group of women workers who have often been considered impossible to organise. Its origins go back to June 1988 when paid workers and homeworkers active with the Leicester Outwork Campaign spoke at a two-day meeting held in Bradford. As a result a small group of women from voluntary organisations in West Yorkshire began to work round the issue.

We have never been in the position of the Leicester Campaign where there is sufficient support for an independent campaign but have had to rely on the resources of existing organisations, in particular the Yorkshire and Humberside Low Pay Unit and Leeds Industrial Mission working in liaison with local authorities. There is a national network of homeworking groups and we take part in a regional network in the north of England so that we can meet regularly and exchange experiences with women working round homeworking. At present, this network meets every two months and includes women from Liverpool, Manchester, Rochdale, Sheffield, West Yorkshire and Newcastle.

Our basic approach has been that contact with homeworkers is the first and essential step. Without this contact being made, we have little knowledge of what is going on, how women feel about their work or what problems they face. We adopted the Homeworkers' Fact Pack – an information pack on employment, welfare benefits and other rights – from the Leicester Outwork Campaign, as our starting point. The Fact Pack is advertised as available free to homeworkers, who can phone or write to us for copies. From this base, we began to form a picture of the type of work being done at home, the rates of pay and other employment conditions and the questions women have about the work. When there is a specific problem we try to give advice and where possible follow this up with a visit.

The availability of a free advice line at the Yorkshire and Humberside Low Pay Unit is crucial. At the unit, priority is given to the advice line so that calls are always taken in office hours and there is someone to talk to.

The *Outworkers' News* was originally a photocopied newsletter which we have 'upgraded' to keep in touch with a growing number of women who have contacted us over time. It is also distributed to community and advice centres so that, in turn, it becomes a way of reaching more women. We encourage homeworkers to write of their experiences in the newsletter, so that by building links, and exchanging information, they feel less isolated.

The newsletter now prints articles in five languages apart from English – Punjabi, Urdu, Gujarati, Bengali and Chinese. This is important if we are to reach the many women in minority communities who are trapped in their own homes by their situation as women but also face the additional difficulties of living in a racist society.

We have to publicise the existence of the group in an ongoing way, using the Fact Pack, free information and advice as a basic contact point. This involves updating and translating material into different languages and dealing with queries as they come up. We also use the local press and radio to reach more people as well as specific campaigns like a stall in Leeds market to launch the Leeds Fact Pack and a Christmas campaign to highlight the work of homeworkers.

The West Yorkshire Homeworking Group documents the extent of homework mainly through direct contact with homeworkers. But other sources of information include newspaper adverts, employers and academic research. Our surveys have revealed that approximately half the work is in the textile and clothing industries. Outmending and burling (inspection and darning) of the finished cloth were always among the most skilled jobs in a mill and have for many years been put out to married women to do at home. It is particularly common in the Bradford and Kirklees area and has increased in recent years as a result of the closing down of mending rooms in the main mill, with the work being subcontracted out. In the carpet industry, sample books are made at home as they are for other forms of woollen cloth and yarn. At the other end of the textile industry, rag picking and sorting is still done at home.

Garment making is probably the most common form of homework found in West Yorkshire, as in other parts of Britain. There is often a chain of subcontracting involved, leading to big retail stores. Some firms have only a core of cutters and finishers, with all the machining being put out to home-sewers. Sometimes this is skilled work of making through the whole garment. In other cases, it consists of a single operation, for example sewing a collar or cuff, or inserting a zip-fastener. The range of garments is wide: children's clothes; trousers; blouses and shirts; dresses; canoeing jackets; sporting wear; medical uniforms, among others. Both fur and leather garments are made at home. Other kinds of sewing include soft toys; velvet hearts for inclusion in cosmetic sets; sewing horse blankets; sewing flashes on to trainers and fine craft sewing, such as quilting, smocking and embroidery. Work with leather done at home includes clothing mentioned above, but also making handbags and other leather bags; sewing up moccasins and sewing or gluing small pieces of suede or leather, either for the manufacture of bags or for sale as window-cleaning or car-polishing rags.

Both hand and machine knitting are found in the home and are among the lowest-paid jobs. The employers vary from small local shops to national mail-order companies. Some of the knitted goods are exported. The big wool manufacturers also employ homeworkers to knit samples for their retail wool outlets.

There is nothing new about engineering work being done at home. One of the largest local engineering firms (now closed down) used to employ women assembly workers in their homes when they left full-time work on account of children. The machine they used in the factory was taken to their home so that they could continue in the same job. Currently a range of engineering assembly work is found at home, including ball-bearings; complete cages for budgerigars, and parts for them; assembling washers and fruit machines. A variety of electrical assembly work is done at home, for example switches, plugs and computer leads, and at least two firms in the area put out the assembly of printed circuit boards to women working at home. Similarly, some plastic assembly work is found at home, for example parts for venetian blinds.

A major source of packing and assembly work in West Yorkshire is the greetings card industry, with associated work such as stationery, and 'fancy goods' such as straws, party masks, Christmas crackers or pencil sharpeners as well as a range of Christmas decorations. Toys and board games, or parts for games, are assembled at home, with for example the 'bagging' of counters and dice. But many other goods are also packed at home from babies' nappies to medical supplies, nails and screws, bath plugs, curtain rails, and tights. One woman told us that she had assembled flower pots with artificial flowers and sand. Others make the cardboard containers for fireworks and string tennis racquets.

There are many other forms of homework that do not fit into a neat category, such as painting decorative porcelain cottages or making decorative mice; preparation of food such as Brussels sprouts for wholesalers. We have also found a range of white-collar work at home, not including professional computer work: addressing and stuffing envelopes; data-inputting; accounts work, and word-processing.

The above list is not comprehensive but gives an idea of the wide range of work done at home. It is difficult to generalise about what types of work are being put out, except to say that most of it is labour-intensive and extremely flexible. There is rarely a written contract and this can result in outright abuse or cheating; non-payment is not an uncommon complaint. Homeworkers make up the most flexible workforce possible for employers: there when labour is required and laid off when there is no further need. In the world of fashion, for example, there is no longer only one style or season a year, but several. Many of the retail stores are competing to keep down their stocks and respond quickly to changes in demand. A workforce of homeworkers enables manufacturers in turn to keep down basic costs such as rent, electricity, or storage and their wage bill, and to respond quickly and flexibly to the demands of the retailer.

It is not uncommon to find extremely low rates of pay, below £1 an hour, and often below 50 pence an hour. In one sample study in 1989, in Kirklees, out of twenty-one women who gave rates of pay, eleven were earning below 50 pence an hour.[6] The kinds of work associated with these very low rates include packing greetings cards, particularly gift tags; packing tights; bagging toys; making soft toys; packing and assembling miscellaneous 'fancy goods'; peeling vegetables; some engineering assembly work and knitting. However, other women are paid closer to factory rates of pay for homework. In electronics work, the management of one factory claims that rates of pay for outworkers are in line with those of workers in the factory. But outworkers are excluded from bonus and holiday payments and they themselves reckon that overall their pay is at the most £2 an hour. A small minority of skilled machinists can actually earn well above the rates of pay they would earn in a clothing factory and are satisfied with the hourly rate of pay. Whatever the rate of pay, most women doing paid work at home are treated less favourably than those working outside. Their work is insecure and they have even fewer rights than other workers, for example in relation to redundancy or maternity rights. They are often called upon to do 'rush jobs', involving weekend and night work.

We aim to develop group work with homeworkers as well as publicity, research and providing advice. For example, after initial contacts with homeworkers in Mixenden, an estate on the edge of Halifax in Calderdale, the homeworkers themselves leafleted the estate and met regularly every week for a year. As well as building a local group, they contacted their MP, spoke at regional and national meetings and generally campaigned round the issue of homeworking. The homeworking officer in Wakefield has also been successful in building local groups which women find supportive and helpful in overcoming their isolation: in the words of one member, Jaswant,

> 'I really get a lot of help from the support group that has been set up. You don't feel isolated. Even now when I'm doing my own business, it's useful. I learnt a lot going to those meetings. It's a day out, and I learn things and find out what's going on.'[7]

In the case of this particular group, many of the women shared a common interest in trying to set up their own business as one way out of badly paid homework. This was in contrast with the Mixenden group, who had been doing extremely low-paid packing work and whose priority was campaigning to expose the employers profiting from their work and for more employment rights for homeworkers.

Bernie, from the Mixenden group, explained that she was prepared to do homework because of her children. This is a common reason for women becoming homeworkers.

> 'I used to make Christmas crackers. All in all, it took roughly three minutes to make one . . . and I was paid one pence per cracker. . . . There was many a time I used to stay up all night so as to get my wage packet at the end of the week, for buying shoes, clothes and sometimes food, for my children. Many people thought I did this for leisure money. But this was not the case. I did this to buy things which were a necessity.'[8]

The women receiving the lowest rates of pay are often those who are most afraid of losing their work because of the need for the extra money, however little, that their work brings in. Many women continue to do this very low-paid homework only as long as they have no other choice. Brenda, for example, a sewing machinist from Huddersfield, said:

> 'I started doing homework seven years ago when I had the little lad at home and I couldn't go out to work. The first job I had was sewing sleepsuits and nighties. . . . The next thing I did was knitting . . . . I never worked out the hourly rate. I think I would have cried if I had. . . . Then I did rag dolls and toy mice . . . I did that work for six to nine months . . . then I decided that I needed a proper job and got this work machining jackets for the firm I'm working for now . . . I earn a good hourly rate with this firm and I'm happy with the work.'[9]

Brenda's current homework, 'a proper job' as she calls it, not only provides a good hourly rate, but is regular and she has been able to form a group of homeworkers.

From the beginning of the campaign in West Yorkshire, we found that methods that we used to contact white women drew little response from the black or Asian communities. Early on in our work we relied on Asian women community workers in Bradford who were in touch with many women through existing groups such as sewing or English classes. One of these women had herself done a small survey of homework in the Pakistani community in Bradford and found sewing homework widespread, with low rates of pay and many problems such as non-payment of wages. We found that the insecurity and fear surrounding homework was even greater among Asian women and they were reluctant to disclose that they were doing homework, even to someone they knew well. This has been confirmed by further work we have been able to do as a result of employing three Asian outreach workers to work with Asian women in Leeds, Kirklees and Calderdale. Once trust has been established, however, it is sometimes easier to find out more information, as women know each other and sometimes work together, particularly in a small community.

Machining garments for the cheaper end of the retail stores, women's, children's, sports or casual clothes, seems to have shifted to the Indian, Pakistani, Kashmiri and Bangladeshi communities. The women are often, although not always, working for a small Asian locally-based firm or an agent of a firm elsewhere. But we know from the labels sewn into the clothes that these smaller firms are supplying major retailers, both high-street stores and mail-order firms. This is different from the pattern among white women machining at home, who are often working for an employer for whom they have worked on-site; a more established firm and one that pays better rates of pay. Machining garments is only one type in a varied pattern of homework among white women, whereas it is far more common among Asian women. We are told that this form of homework is common too among Chinese and Vietnamese women although we have had little direct contact so far.

Other forms of homework, usually the lowest-paid, are also found in the Asian communities: sewing slippers; packing cards; making Christmas crackers; packing nappies; peeling vegetables; assembling nuts and screws, among others. We have found only a few Afro-Caribbean women doing homework, with the exception of outmending in the textile industry, which in recent years has been increasingly subcontracted out from the main mill as a cost-cutting exercise.

Many Asian women, however, are restricted to low-paid homework for most of their working lives as a combined result of discrimination in the labour market and pressures from within their own communities to stay at home. This also results in a higher proportion of Asian women relying on homework as their main source of employment. In some Asian communities, it is unusual to find a woman who has not done homework of one form or another.

There is a strong feeling among many Asian homeworkers that it is wrong to take any action against 'our people', making it difficult for them to enforce their rights if their employer is part of the same community. If we are going to work with Asian homeworkers, it is important for us not only to employ workers from the different Asian communities to overcome barriers of language, culture and religion, but to take an anti-racist stand which takes account of the overall position of these communities in our society. Rather than pinpoint, for example, a small Asian employer, we have to understand the structural position of small businesses in the economy particularly in the clothing industry and look at the whole subcontracting chain, in which a major retailer may be playing off one small company against another in order to reduce prices. We also have to understand that the pressures that keep women at home are not only from the family structure, but from racist attacks on the streets, discrimination in employment and training and other aspects of the all-pervasive racism of our society.

There is still debate on the issue on the relative importance of sexism and racism in the situation of homeworkers in Britain. It is likely that this will take on increasing importance in Europe as a whole as there are significant differences in the understanding of these issues. On a practical level, the Dutch homework support centres have a policy of always employing black workers alongside Dutch women and their material is translated into Turkish and Arabic. However, in many European countries and generally in the European Community at official levels, the question of black women is still analysed in terms of migrant labour and the emphasis is on their particular problems rather than on the racism of the European country and its institutions.

In spite of the differences in the situation of white and Asian homeworkers, and between the different Asian communities, homeworkers face many of the same problems. The whole situation of women doing paid work at home is surrounded by insecurity and fear. At its simplest, most women are afraid of losing their work if they speak to anyone about what they are doing. Others may think they are doing something illegal, working at home. It takes time and patience to win the trust of homeworkers and we have to take a long-term view, and be prepared to respond to the needs of homeworkers, whether these are directly related to the work or not.

Ways of organising

Women's paid work in the home is rarely recognised as valuable employment in its own right, even if it is seen at all. Most of us have had the experience of asking someone whether they know of any homeworkers and getting an immediate negative response. When we go on to explain in more detail what we mean by homeworking, then people remember a relation or friend who used to pack cards, knit sweaters or sew garments in the house. It takes a conscious process to break down the idea of 'work' as being outside and what goes on in the home as not being work, whether domestic unpaid labour or paid employment.

Similarly, homework is not seen as 'real' work, deserving 'real' wages. Knitting, a highly skilled craft, is usually seen as a hobby, done for enjoyment, even when the end product is being exported or sold through high-class mail-order catalogues. One 80-year-old woman we know has been knitting for seventy-two years, since she was a girl of 8, and receives £4 for an intricately patterned sweater. Another woman who has twenty years' experience of 'making through' leather jackets is paid only the legal minimum. Another who made fur garments worth hundreds of pounds was paid less than the legal rate. Homeworkers themselves often undervalue their own work and skills and see their problems as individual and personal. Bringing them into a campaign can overcome their isolation and put their own lives in a wider context.

The reasons why women do homework are linked both to their family situation and personal relationships as well as external factors, such as whether alternative employment is available outside the home. While on the one hand we have to take account of questions like childcare or caring for elderly dependants, we also have to deal with a whole range of issues to do with women's lack of confidence or opposition from within the family to their becoming active in their own right. Counselling on personal matters and consciousness-raising are as much part of the work as negotiation with employers or agents.

A homework campaign has to work at a number of levels, in which the personal interconnects with the political, the family situation with work, lobbying

Parliament with small local meetings. We have to be prepared for a long process of building confidence with some women while others can come forward more readily to take part in the campaign. We have to combine giving advice and support with listening to the views of homeworkers and understanding their situation. The majority of homeworkers are working-class women and we have to build a unity across classes, both among homeworkers, as well as between professionals and homeworkers. In practical terms, the homeworking campaigns have adopted a way of organising that reflects the practice of many women's groups, as well as being influenced by the theory and practice of community work. It aims to bring out the strength of women, more often in small groups with a less formal structure and organisation than in a body such as a union. The attitude to work and skills is not a narrow one and a system of networking operates rather than more hierarchical organisations.

Whereas this form of organising works well at the local level, there is a negative side which anyone familiar with the bewildering number of women's groups in Britain will recognise. It is difficult to build strong national and regional structures which can speak for homeworkers in a united way and centralise important discussions, for example on what kind of legislation is needed. If we are going to have an impact nationally, it is crucial that we build an effective independent women's organisation, based on homeworkers' own priorities, perceptions of what change is needed and strategy for change.

It is too simple to assume that all homeworkers have the same responses to conditions and this often throws up contradictions. In Mixenden, for example, one group of women were so angry about the rates of pay and the way they were treated that they wanted to expose what was going on. Others had little alternative but to continue doing the work and were afraid that any publicity would drive the work away. This example reflects a wider conflict between long-term campaigning or organising work and short-term demands. We have to try and develop ways of meeting some of the immediate needs of homeworkers, or women looking for work at home, while drawing them into a campaign for long-term change.

There is a great variety in the type of grievances we deal with. Women who have worked in a factory and acquired considerable skills know what the rates are for the factory workers and their queries tend to be about overtime rates, holiday pay or redundancy rights. In the clothing industry for instance, we found they were more likely to be aware of the Wages Councils than other women. The main protection for these homeworkers has been statutory provisions, brought in as the Trades Boards at the turn of the century. But we often receive enquiries of a more basic kind which would rarely appear on the agenda of a trade union. 'I haven't been paid for some work. I don't know the name or address of the employer. There's a bloke called Mike who delivers and collects the work in a blue van.' 'Mr X owes me £200 and now he's sent a man to take my machine away. They say if I don't let them have it, they're going to fetch the police in.' 'My employer has brought back some sweaters I knitted. She says they aren't the right size. I've knitted 200 sweaters for her and they've always been all right before. She says she's not going to pay me and she's going to charge me for the wool.' 'I've got 9,000 cards in my kitchen and they've never been to collect them. Do I have to keep them or can I throw them away?' 'When I started the work she said she was going to pay me £40 and now I've delivered it she only wants to pay £7.'

A common complaint from homeworkers is underpayment or straightforward non-payment of wages due. We have adopted a range of tactics in retaliation. In one case, women had been packing gift tags at home. The work was delivered to the home and some of it had never been collected. Another group of women had had to collect their work, queuing up outside a warehouse for hours. When they tried to collect payment, they found the warehouse was closed. Neither of these two groups of women knew who their employer was. We had to become detectives, starting from the remaining gift tags, tracing the producer and subcontractor involved. In this case, we negotiated payment for the women. The fact that some of the gift tags involved were being sold through charity catalogues enabled us to use the body that organises these sales for charities, to put pressure on the producer, who then told us who the subcontractor was.

In another instance of deduction of money from wages, we have supported a woman's complaint through the Small Claims Court; in others we have sought the help of local law centres and Citizens' Advice Bureaux. Homeworkers' lack of written contracts presents problems. One woman took on the packing of felt-tip pens, thinking she could earn some money for Christmas. She was originally told the rate of pay was £4.50 per thousand pens and this was later reduced to £4.50 per thousand packs of pens, cutting the rate by a quarter as there were four pens to every pack. Eventually, with the help of a local law centre, agreement was reached on payment.

Although it is important to look at the relevance of traditional employment rights such as redundancy or maternity rights and establish that homeworkers should be treated as employees, it has been equally important for us to look at other rights not normally associated with work. For example, where there is no written contract, no proof of who the employer is, no payslip or record of work done, the right to hold on to goods until payment is forthcoming may be the only legal right with any relevance to homeworkers. Another query we had was from women who had been making decorative mice. They wanted to know whether they would be breaking any copyright or patent laws if they copied the mice, making alterations in clothing and style, in order to sell them for themselves.

In many cases, we also have to look beyond the immediate employer, who may be at the end of a long subcontracting chain. We adopted this tactic when we found that in Mixenden, women were packing Christmas cards for sale in Woolworths, at a rate of pay which they calculated meant that they could earn not more than 50 pence an hour. The homeworkers concerned wrote to Woolworths but received no response until their MP raised the issue in the House of Commons. Eventually a meeting was fixed with the Managing Director of Woolworths, together with other national representatives of the company and two of the homeworkers, their MP and a member of the Homeworking Group. The company denied that the rate of pay was as low as the women stated but promised to investigate. The meeting attracted local publicity and it was clear that we had considerable 'nuisance value' in the eyes of the big stores.

This is not to argue that traditional negotiating is not important for homeworkers. Where unions negotiate within a factory employing outworkers, it is clearly vital. Unions could also monitor subcontracting chains and raise questions about pay and conditions in smaller firms. But for many homeworkers working for non-union firms, such negotiation is still a long way off and a wider approach is needed.

Another issue frequently raised by homeworkers is health and safety. Two main problems recur; dangerous substances being used without adequate information or protection and the physical and psychological stress caused by long hours of badly-paid work, meeting rush deadlines, while combining paid work with the care of children or other family tasks. A high proportion suffer from different forms of repetitive strain injury, a disease only recently diagnosed as resulting from work and often still wrongly diagnosed by doctors as arthritis or even symptoms of the menopause. One woman, for example, has recently had two major operations on her hands as a result of Carpal Tunnel Syndrome, caused by long hours of sewing homework, over twenty years. If a woman continues her work, she may well risk her hands becoming totally crippled and a great amount of pain. As long as women are working long hours at low-paid work at home, they will continue to be vulnerable to this disease as well as general feelings of depression and stress.

The experience of leather workers in Italy shows that the threat of dangerous substances being used in the home is common to homeworkers in different countries. Leather workers in Naples were using a glue for leather work, making handbags and shoes, which contained a substance that caused polyneuritis, which in its severest form causes paralysis of the whole body. This was exposed after an epidemic of the disease in Naples in 1973, but cases since then have shown that the problem has not been stopped, particularly for workers in the leather industry which employs many women at home.[10] Similarly, Lucita Lazo, who runs an ILO project for homeworkers in the Philippines, Indonesia and Thailand, identified health hazards as a major aspect of homeworking in these countries. She mentioned specifically strain on the eyes of embroidery workers in the Philippines and breathing problems caused by dust from knife-grinding in the home.[11]

Problems arising from homework affect not only the worker herself but also others in the family, including children. One homeworker we talked to said that she gave up knitting mohair sweaters after the family could taste more mohair in their tea than food. A more serious hazard of this work is that we have been told by two women that mohair fluff was the cause of their babies developing asthma. The use of industrial machines or substances, the storage of large amounts of materials or goods may also be a danger to the family.

As with other aspects of working with homeworkers, a flexible approach is needed which combines trade union strengths with a wider approach and an alliance with different groups including community-based occupational health and safety schemes. Publicity is as important as round-the-table negotiations with employers and pressure on retail firms may be more effective than tackling an individual small employer. Whichever tactics we use, it is essential that homeworkers are involved in the process. It is their work that could be stopped as a result of our campaigning and their livelihood that is threatened.

International contacts

From an early stage in our work we made contacts not only with groups in Britain but with organisations in other countries working with homeworkers. The Self-Employed Women's Association was of particular interest to us given the communities of Indian origin in West Yorkshire. Another influential model has been the Homework Support Centres in the Netherlands which were set up after a survey done by the Women's Union in 1979 found more than 100 different kinds of homework being done by its members.

The Women's Union in the Netherlands (Vrouwenbond FNV), originally set up to organise the wives of industrial workers, has become part of the official trade union federation although it does not have negotiating rights with employers. It organises all women, whether doing paid work or not, and women can have a dual membership in the Women's Union and the industrial union concerned. Following the survey on homework, the Women's Union carried out further investigations, including visits to homeworking groups in Britain, and raised the issue with political parties and other trade unions in the Netherlands. Eventually the first homework support centre was set up in the Netherlands, followed later by two others with government funding and one independent centre. The Women's Union acts as a link between the support centres, the government and trade union movement and has ensured that the issue of homeworking is kept in the forefront.

Where possible, we have tried to involve homeworkers in all activities. Karen, an active member of the Mixenden group, attended a conference held in the Netherlands to look at the changing international division of labour in the clothing industry. Another member of that group, Bernie, addressed the European Parliament in Brussels, as part of a delegation organised by the European Network of Women. Elaine, who has done homework for twenty-two years in the Kirklees area, attended a European church conference in Naples and spoke out about homework, as part of the general discussion on the growing divide between rich and poor in the world.

Contacts with the National Support Centre, in Hengelo, led to one of the most successful events that we have organised, a delegation to visit the homework support centres in the Netherlands for which we obtained church funding. We gave priority to homeworkers, or ex-homeworkers, on this visit and of the eight women from West Yorkshire, six were either currently doing homework or had in the past. They came from different areas of West Yorkshire – Calderdale, Wakefield, Leeds and Kirklees – and had experience of sewing, electronics, toy making and the greetings card industry. For many, this was their first trip away from the family and away from England.

The visit was a valuable opportunity to learn about the situation of homeworkers in the Netherlands and their methods of organising there. It also strengthened their confidence as a group and as individuals in their own right. Linda Devereux, a former electronics homeworker who works for the Homeworking Group as a researcher and organiser, reported:

> At the meeting with the Women's Union FNV, I was struck by the chairwoman, Maria. She came across as having such power. She knew all the difficulties and what needed to be done. . . . I felt there was some hope with people like her on our side. This followed through all the way after the meeting . . . I felt more confident. I had seen people become more confident, not afraid to talk up, knowing what to say and with assurance.[12]

The West Yorkshire Homeworking Group is a locally based group. Its main task has to be building contacts with and a network for local homeworkers. But we also need a long-term strategy of where we are going and, with scarce resources, we need clear priorities and an understanding of different strategies, at local, national and international levels. It is here that contacts with other organisations both within Britain and in other countries can provide valuable lessons.

Trade union organising

Our experience in West Yorkshire, which reflects that of other campaigns in Britain, is that few homeworkers are members of unions. There has been a significant shift in national trade union policy on the issue of homeworking, but this has had few practical results in terms of membership although the unions have consistently supported the work of the homeworking groups, particularly at the level of the National Group on Homeworking.

There have been some practical moves recently, particularly in Leicester, where the campaign has been working closely with one of the trade unions and a group of homeworkers has joined the union, with the aim of developing a specific branch for homeworkers. The unions can play an important role in campaigning for legislative change and in providing legal help for fighting test cases under existing law to establish employment rights for homeworkers. However, in general, they have found it difficult to spare the resources needed to work with homeworkers, or at local level there has not been the commitment. If work is being subcontracted out from a union-organised firm, we have found that the immediate union reaction is to stop the work going out in order to protect the jobs and interests of the workers within the factory. Given the difficulty of contacting and organising homeworkers, this is in many ways an understandable reaction. But it hardly benefits homeworkers, who in most cases are afraid of losing their work, if this is what happens when a union is brought in. It is a more difficult process for the union to contact, negotiate for and organise homeworkers who may be working in isolation and scattered over a wide area.

In many cases, there is no union in the companies which employ outworkers. Where this is the case, it is only too easy for an employer to threaten to move the work elsewhere if there is any suggestion of homeworkers' organising for better rates of pay. One group of women explained to us how they had tried to work together to negotiate for better rates of pay for packing cards. They found the employer unwilling to give them any further work. All the homework campaigns find that there are many women desperate for work at home. The threat of moving the work away is real and only a minority of women have worked regularly enough to have any employment rights or to be able to use their skills as a bargaining weapon.

It is a common experience throughout Europe that traditional forms of trade union organising have not been sufficient for organising homeworkers. There are few examples of practical work with homeworkers or traditional methods such as a strike, although there may be more that we do not know about. At a conference in Nantes, France, in 1990 organised by the Centre de Documentation du Mouvement Ouvrier et du Travail, a clothing outworker from Brittany, Eva Sitchenko, told of her experience when all the outworkers at the firm where she worked went on strike. She recounted their grievances, familiar to the experience of women in other countries, of changing piece-rates, lack of payment of expenses and inaccurate recording of their work. Their accumulated grievances led to a strike in 1984 lasting one and a half months. The twenty outworkers held out in solidarity but they failed to win the support of the workers in the factory. A legal case has now been won by the workers, with the support of their union, and significant amounts of compensation have been paid although they have lost their work.[13] This experience in France, where homeworkers have good employment rights at least on paper, is in contrast to the experience of outworkers in Greece. In 1985, homeworkers assembling dolls and other toys, on the island of Zakinthos, went on strike. They lost not only their work which was moved to another part of Greece, but their legal case as labour laws in Greece offer little protection to homeworkers.[14]

Trade unions in Britain have traditionally relied on the economic power of their members rather than legal measures and as homeworkers have little economic bargaining power, it is natural that traditional methods and attitudes have failed them. Whereas in many European countries national collective agreements carry the force of the law, in Britain this is not so. An employer who does not wish to pay his employees, whether factory or outworkers, union-negotiated rates is not obliged to by law unlike in other countries where if a union includes homeworkers in its national negotiations this has at least legal backing. Over the last twenty years, there has, however, been a shift in attitudes and unions now generally recognise the need for a legal minimum wage covering all employees, and increasingly the need for homeworkers to be covered by this and other employment laws.

There are also positive examples of union action, apart from SEWA, where new methods and approaches have been developed, incorporating strategies other than traditional union methods, to work with homeworkers. At the International Labour Organisation experts meeting on homework, held in October 1990, Ela Bhatt (see Chapter 5) from SEWA was one of six 'experts' representing the official union views on homeworking. It was also significant that at this meeting another expert from the Dutch trade union movement, Catelene Passchier, and an observer representing the British Tailor and Garment Workers' Union, Anne Spencer, both expressed the view that the trade unions by themselves cannot organise homeworkers. Drawing on the experience in the Netherlands and Britain, they concluded that there was a need for support centres or homework campaigns to work in alliance with the trade unions.[15] Clearly there cannot be a blueprint for what organisational form this should take. In India, SEWA is formally recognised as a trade union, even though many of its methods are non-traditional particularly when compared with unions in the west. In the Netherlands, the work has taken the form of an alliance, between homework support centres, the Women's Union and the rest of the trade unions.

An example from a paper given by a member of a homework support centre in the Netherlands illustrates how working with homeworkers can then in turn feed back into union negotiations:

> A group of outworkers contacted the outwork project with complaints about the piece-rates . . . and the fact that they were paid only once in three months. The trade union representative was willing to negotiate for them and the employer acknowledged him in that position. This resulted in an increase of the wages, a monthly payment and an allowance for transporting the goods.[16]

In another example, there were many complaints about the work from both Dutch and Turkish outworkers for a particular company:

> Since the company gets its orders from the printing industry, we have got in touch with the printers' unions. The unions will claim the same terms for outworkers as for other workers in the printing industry, according to the regulations in the collective agreement. To avoid risks for outworkers who have complained to us, the trade union is negotiating with the employer without mentioning the outwork project. The consultants to the outwork project keep in touch with the outworkers to inform them about the action and to control the impact.
>
> There are more ways to put pressure on this subcontractor. Since he gets his orders from other firms, members of the employers' councils of those firms could question their employer about contracting out orders to a company which maintains bad working conditions. In the magazine of the printers' union, members are requested to inform the union when their firm is putting out orders to this subcontractor.[17]

The Clothing and Allied Trades Union in Australia has also taken an interesting initiative. In December 1987, the union produced a report on outwork in the Australian clothing industry which pointed out that outwork had grown to such an extent that there were more outworkers than people working in factories. The report outlined the familiar features of exploitation of homeworkers and included a union strategy for combating this. The union had won an agreement, with legal backing, for full employee rights for outworkers, including prescribed working hours for which women had to be paid whether actual work was available or not, and announced its intention to campaign to make sure that these rights were implemented. In 1988, the union also visited Britain, Italy and Switzerland to examine the experience of others working with homeworkers.

The result has been an energetic campaign to inform homeworkers of their rights, a strategy combining legal rights with organising at the grassroots. Legal cases have been taken up side by side with the employment of liaison workers, including ex-outworkers, and widespread publicity through different channels, including local radio and press, employers' organisations and community groups. The union paper *Ragmag* publishes information in a variety of languages, including Turkish, Greek, Chinese, Vietnamese, Spanish, Arabic, Russian, Urdu and Italian. At the international conference held in the Netherlands in 1990 one speaker reported that in one year over 6,000 outworkers had been in contact with the union; membership of the union had greatly increased along with the number of outworkers who had taken their employer to court.

CHANGING THE LAW ON HOMEWORKING

In the past, trade unions and others working with homeworkers have often concluded that they were impossible to organise and that the solution was legislation, either banning homework altogether or giving homeworkers legal protection. The outcome of the Australian campaign will be an important test of whether adopting a two-fold strategy of using the law to strengthen rights at the same time as grassroots organising in the communities concerned, is an effective way of working with homeworkers.

In both France and Italy, new laws have been brought in to protect homeworkers which have not been completely successful. The Council of Europe study, published in 1989, for example concluded that:

> Where the remuneration of homework is concerned, 19th century conditions indubitably prevail to a large extent in most countries of Western Europe, even in certain countries (France, Italy) where regulations on wages have indeed been introduced but are not enforced properly, if at all.[18]

In her study of homework in Italy, Pauline Conroy Jackson links the question of enforcement with organisation. She points out that the Italian law on homeworking is probably as good a law as can be drafted. The problem is one of implementation and organisation:

> Pre-occupied with employment at the so-called point of production, Italian trade unionists have failed, unlike their employers, to observe that the same point of production has moved 'home', that the days of mass meetings and line

> delegates won't work on a street corner in the neighbourhood. Excluded for so long from trade union agendas, many women regard trade union interest in their problems with a cynicism bordering on hostility to outside intrusion onto their women's territory. Italian women homeworkers' experience is a lesson that a law, however well-drafted, will not succeed if not accompanied by a strong grassroots organisation of the women concerned.[19]

The homework campaign in West Yorkshire, along with other campaigns in Britain, has always taken as one of its basic positions the need for employment rights for homeworkers. We have campaigned against the abolition of the Wages Councils which provided a minimum level of protection to homeworkers in the clothing industry, for example. However, the Wage Councils show that it is not sufficient to have a law on the books to secure minimum wages and rights. Of the numerous women we talked to who should have been covered by the wages orders, a small minority only had ever heard of the Wages Councils, knew that they still existed, or had received holiday pay when this was due to them. Only one woman was ever visited by an inspector in her home. We need more discussion, consequently, of what type of legislation is appropriate and what kind of organisation or implementation is needed to ensure that it is effective.

With legislation as with other questions, the key issue is that homeworkers are consulted and play an active part in the discussions about what kind of legislation is appropriate and how it is to be implemented. Homeworkers need to be organised to ensure that they can take advantage of their newly won rights in practice. Without a strategy for organising homeworkers, there is always the danger that legislation is seen as another way of banning homework, by making it uneconomic for employers to continue to use homeworkers. If this happens without major improvements in childcare, women would face the choice of being without work or having to work illegally. Experience in the USA, where certain types of homework have been banned for many years, shows that a ban does not stop homework but drives it underground.

ALTERNATIVE WORK

Many homeworkers have attempted to escape from the trap of low-paid homework by trying other jobs such as party-plan, child-minding, or dressmaking. In some cases, they have had some success in building up their own small business although this usually involves long hours of work and many women have found that they have had insufficient capital to establish their business. Another strategy often put forward by local authorities is for women to set up co-operatives. While both these possibilities, particularly that of the small business, are often tried by women, it is clear that in order to be successful women need a lot of support, backing and training. Without this, they can simply end up earning a marginal living and with their business eventually collapsing.

One experience was a knitting co-operative formed in Newcastle by eight women home-knitters. They faced a lot of competition, as the skills needed for knitting are common among women. They had to take on firms, employing home-knitters on a much bigger scale on low rates of pay, who also had access to better wholesale terms for supplies and expertise in business development, marketing and design.[20]

Other strategies explored by homework campaigns have involved devising training courses appropriate for women, either at home or in outside employment. Leicester Outwork Campaign, for example, developed a training course directly in response to demands from homeworkers, so that they could learn more machining skills and take on better-paid work in the home rather than low-paid packing or assembly work. In West Yorkshire, we have found that there are few training courses that provide both childcare and a training allowance, the two essential features relevant to homeworkers. Although one skilled homeworker from the Homeworking Group is now employed creating such a course, the demands for such training are much greater than present provision and this is clearly an important area where local authorities could be active in supporting homeworkers. Homeworkers from West Yorkshire also found when they visited the Netherlands in September 1990 that the homework support centres there work closely with training centres which help women returning to outside employment after they have been at home with children. The problem for us is that developing training courses ourselves would demand much greater resources than we have available and would inevitably detract from the rest of the work with homeworkers.

LOCAL AUTHORITIES

Homework campaigns in Britain have generally worked in alliance with local authorities, and training is only one area where there is ground for common work. In the 1980s, the national government, far from wanting to look at protection for homeworkers, promoted the ideals of self-employment and the growth of small businesses and was reluctant to distinguish between the homeworker and the self-employed freelance worker or business person. It was the local authorities that supplied much of the funding for work with homeworkers.

The Leicester Outwork Campaign established a model followed by other campaigns where an independent group was funded by the authority to develop contacts with homeworkers. Other authorities have funded workers within the authority, although this can have the drawback of being too ‘official’ for homeworkers. Many women prefer to talk to someone outside the authority, and fears about losing their work, claiming benefits and more recently about the poll tax, make them more reluctant to talk to an official.

In West Yorkshire, we have had considerable interest and support from the local authorities, who fund the Yorkshire and Humberside Low Pay Unit and have found additional money to develop the homework campaign. This has mainly been through economic development or equal opportunity sections within the authority who have found additional money, for example to fund Fact Packs and later for outreach workers to work round homeworking specifically.

Given the lack of official information on homeworking, some local authorities have been interested in research projects in order to find out more about the role played by homeworking in the local economy and establish how local authority services can be more helpful to homeworkers. Since the only way to find out about homeworking is to talk to homeworkers, this research has been a useful way to make contact with women. Media coverage of the findings of such research has in turn been used for further publicity and contacting more women.

Research can establish the extent of homework in the area and the poor conditions of work of homeworkers. This can then be raised within the local authority in order to look at different aspects of a council's policy or services which have an impact on homeworkers' lives. In some areas, for example, housing departments have changed their policy which had previously been effectively to ban homework. In Leicester, the Environmental Health Department has made available grants for safety equipment for outworkers, such as insulation against noise and vibration; lights; good chairs and lockable cupboards to store dangerous substances or tools.

The most important service which the local authorities provide in relation to homeworkers is still childcare. Provision of cheap, good-quality childcare for children of different ages (pre-school and school age) is still probably the major factor that could increase women's choice as to whether to work at home or outside. But in the general climate of the 1990s, in particular the lack of finance at the local level, there is unlikely to be any fundamental change here.

Some local authorities have also considered the use of contract compliance, or conditions attached to grants or loans to businesses, to improve the situation of homeworkers. However, both these strategies became more difficult to implement throughout the 1980s and also have the drawback that unless there is a strong grassroots campaign with homeworkers in the area, they are difficult to monitor and implement.

Local authority backing made it possible to respond to short-term needs while pursuing a broader approach which by increasing women's choice contributed to a long-term strategy. However, as local funding becomes tighter, there are signs that the more general approach will become more difficult and funding will be concentrated on training and enterprise approaches.

BUILDING ALLIANCES

In West Yorkshire, the homeworking campaign has been built round an alliance of homeworkers, women working in voluntary organisations, members of church and community organisations, local authority officers and others, with the support of the trade unions. Asian women's groups and advice or law centres have provided and maintained contacts.

We have also had consistent support from some sections of the church who are concerned with issues of equality and justice. This is important in broadening the base of the homeworking campaign and we have been able to fund some of the outreach work through church trusts. It has also provided some unusual ways of reaching homeworkers. One member of the homeworking group, Dian Leppington, who works for the Industrial Mission (an industry-based church organisation), regularly preaches in local churches. She rarely gives a sermon without talking about homeworking and is usually approached afterwards by at least one member of the congregation with experience of homework.

A concerted and effective campaign requires the organisation of homeworkers along with other forms of pressure. It is important to pinpoint the 'ultimate employer' or the retailer who is selling the goods rather than necessarily focusing on the immediate employer, who may be a small firm desperately competing with other small businesses. For example, a greetings card firm based in Bradford has a network of subsidiaries who produce their cards and organise the mail-order distribution. In the 1980s these firms did not employ homeworkers directly but subcontracted work out to contract packers who then in turn employed homeworkers, many of them at rates of pay that worked out at less than £1 an hour. In trying to improve pay and conditions for these homeworkers, it is more effective to focus on the big company or retail store, and it may be possible to work with some of the smaller employers, other customers of the main firm which in this case includes numerous charities, and with the growing number of consumer pressure groups.

The 'Clean Clothes Campaign' started by Traidcraft and New Consumer aims to make consumers aware of the producers of their clothes and use their buying power to put pressure on retailers. Campaign workers have held discussions with homework campaigns as well as with representatives of workers in the Third World.

A precedent has recently been set by Littlewoods Organisation, a major force in retailing and mail-order catalogues, who brought in a mandatory Code of Practice for their suppliers. Littlewoods seems to be responding to combined pressure from trade unions, women's and consumer groups. At this stage, it is still unclear how effective such a code will be. If it is carefully monitored and properly implemented, it could have a significant impact on homeworkers. It could also present an example of an alliance with one big retailer in order to put pressure on others to follow suit.

Publicity or campaigns can also be linked with the development of alternative trading organisations. Traidcraft's original and main purpose is their mail-order business, marketing products from the Third World, which they guarantee are supplied by businesses and co-operative organisations which pay a fair price to their workers. The implications of this strategy are reflected in SEWA's activities in developing co-operatives, a women's bank and in development issues in general. The issue was also raised by Lucita Lazo, who works with women rural homeworkers in South-east Asia, as a possible strategy to link homeworkers in Europe with those in Asia. Although such initiatives with homeworkers in Britain have not been very successful up to now, we need to develop alternative work as a means of countering the employer's threat to move work away.

At local, national and international level, the homework campaigns need to build alliances with a variety of forces, not necessarily those usually associated with working for better pay and conditions of employment.

There still remains a major problem of women's invisibility or women's representation in major international bodies. Organisations such as the International Labour Organisation and many EEC bodies operate on a tripartite basis, with representation from governments, employers and unions, none of which generally fully reflect women's views, let alone those of homeworkers. Although there has been some change here, particularly with recognition of SEWA by the ILO, grassroots organisations representing homeworkers still face a problem in making their voice heard in such bodies.

A European Working Group was set up in 1992 to develop policy on homeworking for the European Commission. The International Labour Organisation is currently studying the issue of homework. The 'expert' meeting held in November 1990 did not reach agreement on the need for an international convention to set minimum standards in relation to homework but the question of homework is still on its agenda and its regional seminars and study programmes will give those working with homeworkers an opportunity to meet and exchange experiences.

It is vital to complement the work through international organisations by establishing links between the grassroots organisations. The companies which employ homeworkers, or sell the goods produced by homeworkers, in an area like West Yorkshire often themselves operate on a world scale and we have to respond likewise. Homeworkers sewing garments which end up for sale in our retail stores are often at one end of a chain of suppliers which also ends up in factories or with homeworkers in the Third World. A study by the Centre for Research on Multinational Organisations (SOMO), in the Netherlands, showed how one big clothing retailer subcontracted work on the one hand to the Philippines, Thailand and Bangladesh and at the same time to small manufacturers within the Netherlands, who in turn subcontracted out some of the work, often to homeworkers, both Dutch and Turkish.[21] We know that a similar pattern is in operation in Britain. As long as we work in isolation, within national or regional limits, it is possible for the workers in one country to be pitted against those of another in competition for a limited amount of work.

At the international conference held in the Netherlands in May 1990, women spoke about homeworking in several European countries, in Hong Kong, South-east Asia, India and in the refugee camps of the Palestinians. What was striking about these statements was that despite quite distinct conditions and histories, there were common aspects to women's work, particularly homework. Everywhere the work is invisible, undervalued and underpaid, even though it often provides the main income for a family. Everywhere homeworkers have few rights and face common problems. And everywhere they are trying to do the double shift of paid employment and caring for the family.

It is unlikely that these international contacts will come together in a formal organisation. Rather they can operate on the basis of networking, providing a forum for exchanging ideas and experience, for education and solidarity, for building links and practical support.

Most of all, probably, the example of SEWA is useful, not for us to copy their experience, but to learn to build an organisation based on the needs of homeworkers. There are many different strategies open to us and rather than choose between them, we have to develop flexible forms of working, based on the organisation of homeworkers at the grassroots. Different methods can then be adapted in order to combine short- and long-term objectives and build alliances with a range of different forces. In the course of doing this, we can bring together all the strength and skills of women homeworkers, not only to fight for better pay and conditions, but with the overall aim of empowering women in all aspects of their lives. If this sometimes seems a daunting task, the example of those who are further along the road can serve as an inspiration to us. At the May 1990 conference in the Netherlands, the founder of SEWA, Ela Bhatt, said of their work in India: 'When women are organised on the basis of their work, they realise that they are workers and producers. When they work together in a common cause, they forget differences of caste and religion.'[22]

In the homeworking campaign in West Yorkshire, too, women can come together to work in a common cause on the basis that their contribution to the economy needs to be brought out into the open, their voices heard and their work properly rewarded.

NOTES

1 See for example Shelley Pennington and Belinda Westover, *A Hidden Workforce, Homeworkers in England 1850–1985*, Macmillan Education, London, 1989, and Sheila Allen and Carol Wolkowitz, *Homeworking, Myths and Realities*, Macmillan Education, London, 1987.
2 Penny Summerfield, *Women Workers in the Second World War, Production and Patriarchy in Conflict*, Routledge, London 1989, p. 142.
3 Marie Brown, *Sweated Labour: A Study of Homework*, Low Pay Unit, London, 1974.
4 See Swasti Mitter, 'Industrial Restructuring and Manufacturing Homework: Immigrant Women in the UK Clothing Industry', *Capital and Class*, 27, Winter 1986; and Annie Phizacklea, *Unpacking the Fashion Industry*, Routledge, London, 1990.
5 Council of Europe, *The Protection of Persons Working at Home*, Report prepared by the Study Group of the 1987/88 Co-ordinated Social Research Programme, Strasbourg, 1989.
6 West Yorkshire Homeworking Group, *A Penny a Bag, Campaigning on Homework*, Yorkshire and Humberside Low Pay Unit, 1990, pp. 37–9.
7 Jane Tate, interview, August 1990.
8 Jane Tate, interview, August 1990.
9 Jane Tate, interview, August 1990.
10 Pauline Conroy Jackson, 'Italian Homeworkers Submerged in the Economy', unpublished paper, February 1988.
11 International Restructuring Education Network Europe (IRENE), 'Not a Proper Job', Report on the International Conference on Homeworking, *IRENE Newsletter, 12*, September 1990, p. 21.
12 West Yorkshire Homeworking Group, *Outworkers' News*, 5, January 1991, p. 9.

13 Jane Tate, Report on International Conference on Homeworking, organised by IRENE, May 1990, MSS notes.
14 Jane Tate, Report on Conference on Homeworking, organised by Centre de Documentation du Mouvement Ouvrier et du Travail in Nantes, France, November 1990, MSS notes.
15 International Labour Organisation, *Social Protection of Homeworkers*, Documents of the Meeting of Experts on the Social Protection of Homeworkers Geneva, 1990, International Labour Organisation, Geneva, 1991.
16 Chris D'Have, 'Flexible Work and its Implications for Women: Organisation and Strategy through Community/Outwork Projects', Workshop paper, March 1988, unpublished, p. 4.
17 ibid., p. 6.
18 Council of Europe, *The Protection of Persons Working at Home*, p. 55.
19 Pauline Conroy Jackson, 'Beyond Benetton: Homeworking Italian Style', *Yorkshire and Humberside Low Pay Unit Newsletter*, 10, July 1990, p. 15.
20 Jane Tate, Report on Workshop on Co-operative Development, Wakefield, 1989, MSS notes.
21 Marijke Smit and Lorrette Jongejans, 'C&A: The Silent Giant, From Garment Retail Multinational to Female Homeworker', Centre for Research on Multinational Corporations (SOMO), 1989, English summary, unpublished.
22 Jane Tate, Report on International Conference on Homeworking, organised by IRENE, May 1990, MSS notes.

Conclusion

Sheila Rowbotham and Swasti Mitter

Each contribution in *Dignity and Daily Bread* reveals both the extent of the obstacles faced by women workers in employment and the innovatory organisational initiatives which they have taken to counteract them.

Both historically and in the present, a convergence of factors leaves women workers particularly vulnerable. Gender inequality combines with inequalities of race and class. Overcoming these positions of weakness has never been an easy matter even for a workforce that is formally organised, for, as Radha Kumar shows, gender inequalities have been persistently present in trade unions. In the context of new ways of organising, Kumudhini Rosa describes a dilemma which faces the Free Trade Zone (FTZ) women. As a workforce, women's main attraction to investors has been that they are cheap and supposedly docile. Being organised, they thus court the risk of multinational capital simply moving on to countries and sites with no history of labour organisation.

It became apparent during the 1980s that increased vulnerability to capital is not simply a problem facing newly industrialised FTZ women workers, for labour-intensive sweated forms of production were arising amidst capital intensive technological development. The incongruity is particularly apparent in the clothing industry where new technology, as Silvia Tirado explains, gets applied selectively.

The desperation of low-paid workers without highly marketable skills crowding into a sweated trade is not just a contemporary phenomenon, as Sheila Rowbotham shows. However, it arises now at a new conjuncture in which the relation between the state, labour and the market is in the throes of a fundamental realignment.

In the early twentieth century in Britain, politicians and large employers became convinced that a state-regulated, officially organised workforce was not only ethically preferable but likely to be more competitive and socially less volatile. There was an acceptance of a new phase in industrial development by large capitalists like Mond and Cadbury and by politicians and civil servants like Churchill and Beveridge. Likewise, many social reformers and feminists became convinced that it would be better to reduce the numbers of the sweated poor in paid work and safeguard the health of mothers by extending state welfare. This approach to poverty formed the ideological underpinning to the post-war welfare state in Britain. Some of these assumptions about the regulation of women workers also surfaced in colonial India, after undergoing certain bizarre permutations in the process of being transplanted from one social context into another.

A feature of recent economic change has been the disintegration of these ideological certainties of the early twentieth century which became the basis of policy making in countries like Britain. As Jane Tate demonstrates, one consequence of the current reliance on flexible workers by the management in the private and the public sector has been the growing number of homeworkers who are outside the organised trade union movement and the regulated economy.

Not only gender but also class, ethnicity and race determine who is most likely to take on casualised work. In other words, existing economic and social vulnerability gets accentuated as the casualisation of the workforce becomes more common.

The interlocking aspects of women's subordination in society has always presented strategic problems. Women's position in the sexual division of labour and the effects of state policy upon reproduction as well as production, as Sheila Rowbotham notes, often led to linking the casualised women workers with welfare programmes. This strategy characterised the late nineteenth- and early twentieth-century campaigns against sweating in Britain. It was to have an impact in the US rather later during the New Deal, when the Roosevelt government was prepared to listen to reformers who sought to combine protective legislation with welfare. It still has a resonance even though it has become harder to apply because of the international mobility of capital.

As Swasti Mitter points out, the differing character of governments and laws can be of crucial significance for vulnerable workers. However, laws and policies on their own are not sufficient. Alone, legislation can be evaded, in the case of homework and sweated industries. Without the continuing pressure of workers themselves, legislation can also neglect problems and even have consequences which make women's vulnerable position as workers even more oppressive.

Alliances are crucial aspects of poor women's organising efforts. Women workers have rarely been able to exert influence over state policies alone. In contrast, middle-class social reformers have often been able to have a more direct impact on the state. Historically the alliance between middle-class women and working-class women has proved effective when the former have been responsive to the actual demands expressed by the workers. Support from male workers has also been an important factor in the unionisation of women. The danger of protective legislation being used to drive women out of the labour force indicates that there are strong strategic reasons to extend protection to men, rather than emphasising absolute differences between women and men.

The nature of alliances and the manner in which pressure is exerted on the state have been part of poor women's organisational history and they have a contemporary significance.[1] For instance the Women's Trade Union League both in Britain and in the US organised workplaces and put pressure on the state in the late nineteenth and early twentieth centuries. The Nineteenth of September Union in Mexico City followed a similar pattern in the 1980s.

The state, however, cannot necessarily be relied upon to mitigate the harsh effects of the market. A major difficulty of state regulation is that state policies tend to rigidify in ways which can have effects that are contrary to their original purpose. In addition, as Swasti Mitter points out, the state too can have an ambiguity of intention. Not only can it protect the weak, but it can itself contribute to the erosion of workers' legal rights and social gains. In a similar vein, Diane Elson, in 'Structural Adjustment: Its Effect on Women', has observed that it is misleading to simply pose the state against the market, for the state can likewise contribute to the 'social, economic and ideological processes that subordinate women'.[2] Because apparently neutral policies and services such as the provision of water supplies, electricity, waste disposal, transport, health care and education all have gendered implications, Elson states that the aim thus should be 'restructuring both the public sector and the private sector to make them more responsive to women's needs as producers and reproducers'.[3]

The key issue is the extent to which poor women are able to gain greater control over their working conditions and more democratic power over state policies. It has always been difficult for unprivileged groups to define the terms of economic regulation through the state. The aim of alliances should be to facilitate the process. Democratic participation provides a means of flexibly determining needs which can avoid some of the costly rigidities of top-down policy making. But flexibility itself is not a neutral concept. As Swasti Mitter observes, Hernando de Soto's insights into the necessity of flexible forms are valid up to a point. But the question is: who has the power to decide what kind of flexibility?

A recurrent theme in this anthology is that both economic regulation and flexibility should be assessed in terms of their impact on poor women's lives rather than one or the other being presented as inherently preferable. A contradictory requirement of capital has been the need for a regulated workforce and the desire to maximise flexibility mainly on its terms.

Regulation and flexibility thus have pros and cons for employers; the same is true also for workers – especially women workers with children. Aili Mari Tripp describes women seeking greater flexibility in employment and gaining a degree of control amidst economic disintegration in Tanzania, while Jane Tate tells of a growing casualisation process, in which an imposed flexibility means acute vulnerability, even within a western capitalist context.

All the contributions in *Dignity and Daily Bread* suggest a shift in the relation between what are regarded as economic and social spheres of action. In accounts of women's role in social movements the significance of the personal is frequently noted. Indeed Aili Mari Tripp shows how social and community networks among women are assuming an economic role in Tanzania. In focusing on women as workers, interconnections between production and reproduction; sex, mothering, consumption and community, constantly recur. Taking a step of resistance, or seeking a different way of earning a livelihood, thus affects all aspects of women's universe.

Renana Jhabvala suggests that poor women's position in society can be turned from an overwhelming disadvantage into a means of offsetting weaknesses – by pooling strengths acquired in an indefatigable and resilient experience of surviving against the odds. She demonstrates how SEWA has developed a complex and sophisticated organisational theory and practice which combines mobilisation of workers with provisions for social services. While putting pressures on the state it tries to get planning decisions, financial institutions and laws changed; but it also acts as a watchdog making sure that the existing machinery actually works for poor and unprivileged women.[4]

This particular combination of various forms of organisation has proved more successful in mobilising large numbers of poor women than the attempt to unionise in more conventional ways as adopted by the Nineteenth of September Union. Casualised workers who are not accustomed to organisation often become disillusioned when a union fails to accomplish dramatic change. As Silvia Tirado explains, that has been one reason for the decline in numbers in the Nineteenth of September Union. The provision of practical services can be a crucial factor in making participation continuous. The relative success of SEWA's strategy of combining several kinds of organisational forms, however, is partly attributed to state support which the Nineteenth of September Union did not have.

SEWA itself is not without its own problems. In avoiding some organisational blockages, SEWA has encountered others. For example SEWA faces a contradiction which is comparable to Emma Paterson's Women's Trade Union League in its early phase, as described by Sheila Rowbotham. SEWA's social services and the fostering of morale and confidence involve its dependence on outside funding, whereas an important aspect of trade unions is their autonomy. The very success of SEWA too has led to difficulties. As it has raised the confidence of the women, their expectations have altered. For example, when paper pickers were organised by SEWA into a cleaners' co-op, another union came along and organised them to bargain for better pay. As Sallie Westwood has observed, 'SEWA members had learned some trade union principles early on and they held fast to them'.[5]

It is not clear how applicable SEWA's approach might be in other contexts. Lifting schemes from one context into another is much more complicated than it appears on the surface. Bombay, Ahmedabad and West Yorkshire have long traditions of working-class organising. This is not true for Free Trade Zones of the Third World. Tanzania again raises a set of quite different questions as private enterprise arises out of state socialism. Despite these diversities in industrial cultures, there are pressing reasons for international exchanges of organising experiences. These exchanges could be rewarding, for example, in assessing an alternative course for the small entrepreneurs, described by Aili Mari Tripp, which might avoid the sweated aspect of subcontracted work.

During the 1980s, various kinds of grassroots international networks have developed, linking workers, producers, consumers and researchers. If there is to be cross-fertilisation between various forms of democratic economic innovation, a deeper process of direct communication among the women affected by casualisation is necessary. Jane Tate describes how these direct exchanges have been emerging. They need to be supported by theoretical work which investigates and assesses their potentials and their problems.[6] For the poor women, whose livelihoods are at stake, decisions about whether to join a co-op or participate in a protest in an FTZ or take out a loan to set up a small business at home, are of crucial import. Part of the inequality they face is their lack of time and access to resources for economic decision making. Mistakes cost them and their families dear.

Without falsely idealising organising which is both fragile and arising in adverse circumstances, a creative process is evident in society – a new kind of social and economic democratic practice is laboriously being conceived. Against all odds, utilising predominantly their everyday skills and resources, poor women in many parts of the world can be observed shifting organisational structures and ways of seeing. Dignity and daily bread are very basic demands. Yet these claims upon development, made by conscious human agents who labour for such small rewards, present a tremendous human challenge. What kind of development? What kind of growth? What kind of society can ensure that their aspirations are met?

NOTES

1 For a discussion of the implications of state protection, see Alice Kessler-Harris, 'Protection for Women: Trade Unions and Labour Laws', in Wendy Chavkin (ed.) *Double Exposure, Women's Health Hazards, On the Job and At Home*, Monthly Review Press, New York, 1981; Alena Heitlinger, 'Maternity Leaves, Protective Legislation and Sex Equality: Eastern European and Canadian Perspectives', in Heather Jon Maroney and Meg Luxton (eds) *Feminism and Political Economy, Women's Work, Women's Struggles*, Methuen, Toronto, 1987; P.H. Rohini, 'Women Workers in Manufacturing Industry in India: Problems and Possibilities', in Haleh Afshar (ed.) *Women, Development and Survival in the Third World*, Longman, London, 1991.

2 Diane Elson, 'Structural Adjustment: Its Effect on Women', in Tina Wallace with Candida March (eds) *Changing Perceptions, Writings on Gender and Development*, Oxfam, Oxford, 1991, p. 42.

3 ibid., p. 43. On the impact of structural adjustment on Nigerian women, see Carolyne Dennis, 'Constructing a "Career" Under Conditions of Economic Crisis and Structural Adjustment: The Survival Strategies of Nigerian Women', in Afshar, *Women, Development and Survival in the Third World*.

4 On the male attitudes to women's work in the Textile Labour Association and SEWA's success and problems in combining unionising with co-operation, see Sallie Westwood, 'Gender and the Politics of Production in India', in Afshar, *Women, Development and Survival in the Third World*. On the strengths and weaknesses of the financial strategies adopted by SEWA, Annapurna and the Working Women's Forum, see Jana Everett and Mira Savara, 'Institutional Credit as a Strategy Towards Self-reliance for Petty Commodity Producers in India: A Critical Evaluation', in Afshar, *Women, Development and Survival in the Third World*. For a fuller narrative history, see Kalima Rose, *Where Women are Leaders, The SEWA Movement in India*, Zed, London, 1992.

5 Westwood, 'Gender and the Politics of Production in India', p. 302.

6 For examples of accounts of various forms of economic organisation, see Maria Mies, *The Lace Makers of Narsapur, Indian Housewives Produce for the World Market*, Zed Press, London, 1982; Marilee Karl and Choi Wan Cheung, 'Resistance, Strikes and Strategies', in Wendy Chapkis and Cynthia Enloe, *Of Common Cloth, Women in the Global Textile Industry*, Transnational Institute, Amsterdam and Washington, 1983; Jana Everett and Mira Savara, 'Institutional Credit as a Strategy Toward Self-reliance for Petty Commodity Producers in India: A Critical Evaluation', in Afshar (ed.) *Women, Development and Survival in the Third World*.

Name index

Subject index